"Tesla really understood the interconnectivity of the universe."
—Ethan Hawke

"*Inspiration*—that is what he gave to other inventors whose endeavors his life spanned, and that is what his work continues to give to technical specialists in these times."
—Margaret Cheney, *Tesla: Man Out of Time*

"Today, we yet find that the writings of Tesla retain their undiminished power of inspirational endeavor to the reader."
—Leland Anderson

"It is my opinion that he was incredibly ahead of his time."
—Preston B. Nichols, *The Montauk Project: Experiments in Time*

"An eccentric genius who out-invented Edison and discovered radio before Marconi . . . Nikola Tesla may be one of the most important men of invention in our nation's history."
—David J. Kent, *Tesla: The Wizard of Electricity*

"Were we to seize and to eliminate the results of Mr. Tesla's work, the wheels of industry would cease to turn, our electric cars and trains would stop, our towns would be dark, our mills would be dead and idle."
—Bernard A. Behrend

"In almost every step of progress in electrical power engineering, as well as in radio, we can trace the spark of thought back to Nikola Tesla. There are few indeed who in their lifetime see realization of such a far-flung imagination."
—Ernst Alexanderson

"Nikola Tesla, in the opinion of authorities, today is conceded to be the greatest inventor of all times. Tesla has more original inventions to his credit than any other man in history. He is considered greater than Archimedes, Faraday, or Edison. His basic, as well as revolutionary, discoveries for sheer audacity have no equal in the annals of the world. His master mind is easily one of the seven wonders of the intellectual world."
—Hugo Gernsback, *Electrical Experimenter*

My belief is firm in a law of compensation. The true rewards are ever in proportion to the labor and sacrifices made.

HEATHEN EDITIONS
THEIR BOOKS. OUR WAY.

Published in the good ole United States of America
by Heathen Editions, an imprint of
Heathen Creative
P.O. Box 588
Point Pleasant, WV 25550-0588

Heathen Editions are available at quantity discounts.
For information and other tomfoolery, check us out online:

heatheneditions.com

@heatheneditions
#heathenedition

First serialized in *Electrical Experimenter* 1919
Heathen Edition published 2022

Heathen logo, colophon, design, Heathenry, and footnotes
Copyright © Heathen Creative, LLC 2022

All rights reserved.

Book and cover design by Sheridan Cleland
Set in 9pt Droid Serif
Numerals in 36pt Afton James
Titles in 30pt Tesla

ISBN: 978-1-948316-03-3

FIRST HEATHEN EDITION
1 2 3 4 5 6 7 8 9 0

By special arrangement the *Electrical Experimenter* will, beginning with the next issue, publish a series of articles entitled "My Inventions," by Nikola Tesla. The great inventor will contribute a signed article monthly, which articles will run for several years. Most of this material has never appeared in print before. The articles will be published in book form later. We consider this announcement the most important we ever made.

—The Publishers.
Electrical Experimenter
January 1919

An interesting study of the great inventor, contemplating the glass bulb of his famous wireless light. This is the only profile photograph of Mr. Tesla in existence. It was taken specially for the *Electrical Experimenter*.

Contents

Heathenry: Thoughts on the Text ... ix
Nikola Tesla and His Inventions: An Announcement ... xiii
Nikola Tesla: The Man ... xxiii

MY INVENTIONS
I   My Early Life ... 1
II My First Efforts at Invention ... 15
III My Later Endeavors ... 29
IV The Discovery of the Tesla Coil and Transformer ... 45
V   The Magnifying Transmitter ... 63
VI The Art of Telautomatics ... 83

& other essays
The Effect of Statics on Wireless Transmissions ... 107
Famous Scientific Illusions ... 113
Editorial: The New Wireless ... 135
The True Wireless ... 139
The Moon's Rotation ... 161
The Moon's Rotation (Follow-Up) ... 171
Electrical Oscillators ... 187

+ appendices
Tesla Has New Pointless Lightning Rod ... 205
Tesla's Egg of Columbus ... 211

HEATHENRY
Thoughts on the Text

First and foremost, I want to thank those of you who have been patiently awaiting the publication of this particular Heathen Edition. When we Heathens originally planned a tentative, pre-launch release order of Heathen Editions way back in 2016-ish, this book landed in the #3 slot and never moved, and was never going to move, because if you know you know. And of all the Heathen Editions that we've published so far, this one is by far the most image-heavy (84 images/illustrations in total), which drastically slowed its production. Why? Because most publishers who have published a version of this book *do not* include the images, but given how often Tesla makes reference to the illustrations we couldn't imagine publishing our edition without them, and we weren't happy with the quality of the images as sourced from the original *Electrical Experimenter* magazines[1] and also couldn't imagine republishing them as they were (probably why most other publishers opt out), so we set about restoring each of them to a level of quality that we think actually exceeds the as-printed originals. A daunting, laborious task whose fruits we hope speak for themselves.

---
[1] *The Electrical Experimenter*, the successor to *Modern Electrics*, was an American technical science magazine published from May 1913 to July 1920.

Moreover, when we revisited the original *EE* magazines, we took notice of the other articles that Tesla had penned in addition to his serialized autobiography *My Inventions* and wondered why no one had yet collected those in one volume — and that's when it occurred to us: why not collect ***all*** of the articles that Tesla contributed to *EE* in 1919 in a single volume, especially since Tesla cross-references all of the articles so often?

So that, dear reader, is what you now hold in your hands: every article that Tesla wrote and contributed to *Electrical Experimenter* in 1919, including two additional articles (one is actually from late 1918, but Tesla references it so often we felt its inclusion was imperative) that also focus on his experiments and inventions, which may or may not have been authored by Tesla himself — but were more likely penned by *EE* editor Hugo Gernsback.[2]

Gernsback, it would seem, was quite the Tesla devotee, as evidenced by his introductions to each chapter of the *My Inventions* serial, and the other essays as well. From the vantage point of today, some would lead you to believe that Tesla only recently gained respect for his achievements, so we felt it important to include these introductions to better illustrate the reverence that Tesla was afforded by a few in his own time. Like the images, inclusion of Gernsback's introductions also seems to be something no other publisher has yet done until now.

As for the text, we have chosen to modify some of Tesla's odd spelling tendencies for modern eyes (e.g. "tho" is now though, "imprest" is now impressed, and so on). Additionally, we have researched and appended 156 footnotes to better clarify and identify some of the science, technology, and persons noted by Tesla, as well as to help you quickly locate the pages, illustrations, or chapters that he

---

[2] Hugo Gernsback (1884–1967) was a Luxembourgish–American inventor, writer, editor, and magazine publisher, best known for publications including the first science fiction magazine. His contributions to the genre as a publisher were so significant that, along with the novelists H. G. Wells and Jules Verne, he is sometimes called "The Father of Science Fiction."

Thoughts on the Text   xi

cross-references throughout. Where appropriate, we have included text and/or images that Tesla referenced outside of the original *EE* magazines. All to say that we have labored to make an informative and educational edition that we hope is an illuminating read.

Switching gears — if you're a fan of Netflix's series *Stranger Things* (because who isn't?), then you may find this bit of trivia interesting: We recently read the book *The Montauk Project: Experiments in Time*, which is cited as a source of inspiration for the series (Montauk was even the early working title of the show), wherein the author claims that some of the technology used for the Montauk Project was designed for RCA in the 1930s by Tesla using the alias "N. Terbo."[3] For those unfamiliar, the Montauk Project is "officially" labeled a conspiracy theory that alleges there were a series of United States government projects conducted at Montauk Air Force Station in Montauk, New York, for the purpose of developing psychological warfare techniques, which included time travel.

While that may certainly sound far-fetched at the moment, you may not think so after reading this book as Tesla quite specifically outlines that he had developed the technology and *used it* to modify the weather *before* he penned these articles[4] — now over a century ago! — and he also outlines how he had developed "aerial machines devoid of sustaining planes, ailerons, propellers and other external attachments, which will be capable of immense speeds."[5] You know, like UFOs? One may be able to argue how those words can be interpreted, exactly, but it's far more difficult to misinterpret the Tesla-approved illustration featured on p. 102.

And so, here you have Tesla confirming — again, *over a century ago* — two *science facts* that most people today still claim to be in the realm of *science fiction*. So, when it comes to a Tesla time travel

[3] Nichols, Preston B. (1992). The Montauk Chair. *The Montauk Project: Experiments in Time* (p. 70). Sky Books.
[4] See p. 67–68; p. 86
[5] See p. 103

ch. ends next p.

allegation, it may be worth noting that *The Montauk Project* author Preston B. Nichols said that Tesla was "ahead of his time." And Margaret Cheney subtitled her Tesla biography "Man Out of Time."

Maybe they're trying to tell us something?

As for us Heathens, we love that there are far more questions than answers when it comes to Tesla——

May that mystery live on!

NIKOLA TESLA
AND HIS INVENTIONS
An Announcement

Several years ago, in the course of a discussion, a well-known journalist asked me whom I considered at present the world's greatest inventor. I said: "If you mean the man who really invented, in other words, originated and discovered not merely improved what had already been invented by others, then without a shade of doubt, Nikola Tesla is the world's greatest inventor, not only at present, but in all history."

My friend was much surprised and voiced his astonishment. "Surely," said he, "you do not mean to place Tesla ahead of such great men as Archimedes, Faraday, or Edison?"

"That is exactly what I mean," I replied, "and before twenty-five years have elapsed the world at large will echo my opinion."

"But listen," persisted my friend, "who on earth is this man Tesla anyway? What are his wonderful inventions, what great thing has he ever done? How is it that the world at large does not know him?"

"To begin with, and the better to impress you," I replied, "Tesla has secured more than one hundred patents on inventions, many of which have proved revolutionary. Science accords to him over 75 original *discoveries*, not mere mechanical improvements. Tesla is an originator in the sense that Faraday[1] was an originator. Like the latter his is a pioneer blazing the trail; aside from this he is a discoverer of the very highest order.

"Ninety percent of the entire electrical industry pays tribute to his genius. All electrical machinery using or generating alternating current is due to Tesla. High tension current transmission without which our long distance trolley cars, our electrified lines, our subways would be impossible, are due to the genius of Tesla. The Tesla Induction Motor, the Tesla Rotary Converter, the Tesla Phase System of Power Transmission, the Tesla Steam and Gas Turbine, and the Tesla Coil and Oscillation Transformer are perhaps his better known inventions.

"As to your last question, namely, why the world at large does not know Tesla, it is answered best by stating that he has committed the unpardonable crime of not having a permanent press agent to shout his greatness from the housetops. Then, too, most of Tesla's inventions, at least to the public mind, are more or less intangible on account of the fact that they are very technical and, therefore, do not catch the popular imagination, as, for instance, wireless, the X-ray, the airplane, or the telephone."

The trouble with Nikola Tesla is that he lives a century ahead of his time. He has often been denounced as a dreamer even by well informed men. He has been called crazy by others who ought to know better. For Tesla talks in a language that most of us do not as yet understand. But as the years roll on, Science more and more appreciates his greatness, and begins to pay him tribute more and more.

[1] Michael Faraday (1791–1867) was a British scientist who contributed to the study of electromagnetism and electrochemistry. His main discoveries include the principles underlying electromagnetic induction, diamagnetism, and electrolysis. He also created what is known today as a Faraday cage (an enclosure used to block electromagnetic fields).

An Announcement

In 1893, three years prior to the earliest attempts in Hertz[2] wave telegraphy, Tesla first described his wireless system and took out patents on a number of novel devices which were then but imperfectly understood. Even the electrical world at large laughed at these patents. But large wireless interests had to pay him tribute in the form of real money, because his "fool" patents were recognized to be fundamental. He actually antedated[3] every important wireless invention.

A few weeks ago the world read through news dispatches of a great wireless discovery — the static eliminator. But Tesla had not only patented systems overcoming this and other forms of interference but had actually constructed and successfully operated devices years ago in Colorado, under conditions where static interference was troublesome to an extraordinary degree. A photograph of one form of his apparatus is published with a note from him for the first time elsewhere in this issue of *Electrical Experimenter*.[4] And so it goes. The world smiles an unbelieving smile, but Tesla's master mind invariably sets the world aright.

I first read about Tesla in a well-known German weekly publication when I was less than 15-years-old. The Editor of that publication reproduced his picture on a full page and paid high tribute to Tesla, hailing him as the world's coming greatest electrician.

H.W. Buck, Chief Engineer, President of the American Institute of Electrical Engineers,[5] among others, said: "The work of Nikola Tesla in his great conception of his rotary field seems to me one of

---
[2] Heinrich Rudolf Hertz (1857–1894) was a German physicist who first conclusively proved the existence of the electromagnetic waves predicted by James Clerk Maxwell's equations of electromagnetism. The unit of frequency "cycle per second" was named the "hertz" in his honor.
[3] Precede in time; come before (something) in date.
[4] See "The Effect of Statics on Wireless Transmission" (pp. 107–111).
[5] The American Institute of Electrical Engineers (AIEE) was a United States-based organization of electrical engineers that existed from 1884 through 1962. On January 1, 1963, it merged with the Institute of Radio Engineers (IRE) to form the Institute of Electrical and Electronics Engineers (IEEE).

the greatest feats of imagination which has ever been attained by human mind."

Lord Kelvin,[6] before the British Association, commenting upon the Tesla Transformer exhibited, said: "This is a wonderful development of the induction coil destined to be of great importance."

Electrical Review,[7] commenting upon the wireless: "Mr. Tesla's researches in this field have attracted worldwide attention, and his is undoubtedly the master mind."

Der Electro-Technische Anzeiger, Berlin, and *Elektrizität*, Leipzig, Germany, (commenting upon Tesla's work): "It is a combination of the grandest power of technical performance with the most vivid imagination, such as has never been manifested itself in the human mind."

Brigadier Allen, of the United States War Department (commenting upon Tesla's Turbine): "Something new in the world. Officers are greatly impressed with it."

While studying abroad I read every scrap of his work I could lay my hands on. I performed most of his high frequency experiments, and the more I saw of his work the more impressed I became. Some years ago as Editor of *Modern Electrics*,[8] I met him in a New York shop where his famous turbine models were first built. I was fascinated with the tall, gaunt man, then about 50 years old, but looking less than 30. His extraordinary face, with his deep-set blue eyes proclaimed the intense thinker — the philosopher. A few minutes' chat with him left me more than ever convinced of his greatness.

[6] William Thomson, 1st Baron Kelvin (1824–1907) was a British mathematical physicist and engineer who did important work in the mathematical analysis of electricity and formulation of the first and second laws of thermodynamics, and did much to unify the emerging discipline of physics in its modern form. Absolute temperatures are stated in units of kelvin in his honor.

[7] *Electrical Review* is the longest established UK electrical journal, first published in 1872 as the *Telegraphic Journal*.

[8] *Modern Electrics*, created by Gernsback in 1908, was a technical magazine for the amateur radio experimenter and was published monthly until 1914.

An Announcement

Further contacts during the past few years still enhanced my opinion of him. Tesla is a man of extraordinary knowledge. He is remarkably well read and has a photographic memory whereby it is possible for him to recite page after page of nearly every classical work, be it Goethe, Voltaire, or Shakespeare. He speaks and writes twelves languages. He is an accomplished calculator, who has little use for tables and textbooks and holds the sliding rule in contempt. Tesla has received numerous honors and distinctions of all kinds. He is a knight of several orders, holder of many titles and diplomas. Some time ago he was awarded the Elliott Cresson gold medal[9] by the Franklin Institute[10] and last year the Edison medal by the American Institute of Electrical Engineers. Many extraordinary distinctions have been offered to him which he has declined. As of timely interest one instance may be mentioned. At the announcement of Tesla's high frequency discoveries, while the former Emperor of Germany[11] was all-powerful and great men were eager of his favors, Tesla received an invitation from him and the Empress to repeat his celebrated experiments at the Royal Palace in Berlin. He forgot all about it and did not answer for one year, when he politely apologized for his inability to avail himself of the honor. Later the invitation was renewed and nearly two years passed before Tesla answered to the same effect. After a lapse of time, however, upon

[9] Tesla received the award in 1894, which was the highest award given by the Franklin Institute "for some discovery in the Arts and Sciences, or for the invention or improvement of some useful machine, or for some new process or combination of materials in manufactures, or for ingenuity skill or perfection in workmanship." The Institute continued awarding the medal on an occasional basis until 1998 when they reorganized their endowed awards under one umbrella: The Benjamin Franklin Awards.
[10] Founded in Philadelphia, Pennsylvania, in 1824, and named after Benjamin Franklin (1706–1790), the Franklin Institute is one of the oldest centers of science education and development in the United States and houses the Benjamin Franklin National Memorial.
[11] Wilhelm II (Friedrich Wilhelm Viktor Albert; 1859–1941) was the last German Emperor (Kaiser) and King of Prussia, reigning from June 15, 1888, until his abdication on November 9, 1918.

the announcement of another important invention he received the invitation for the third time, with the assurance that an altogether unusual honor was reserved for him. "Well, boys," said Tesla to his assistants after he laid the invitation which he never answered aside, "the Emperor must be a great man. I do not think that I would be capable of acting in this way if I were in his place." Perhaps the most remarkable tribute was paid to him when he made his famous experiments in Colorado in 1899. It was by J. Pierpont Morgan,[12] the elder, who donated $150,000, which enabled Tesla to produce artificial lightning and incidentally to electrify the entire earth.

Some of Tesla's inventions have been of far-reaching importance in the War. The resources and productive powers of the country have been greatly increased through extended use of his system of alternating current transmission and transformation of energy. Nearly ten million horsepower of water falls have been harnessed by this means, thus saving forty percent of the entire coal output of the United States. The railroads have been electrified and his induction motor has revolutionized the steel industry and operation of factories. His electric drive has been adopted on the largest cruisers and battleships as the most perfect means of propulsion. His wireless inventions have proved indispensable and his oscillatory apparatus has been of inestimable service in chirurgical[13] and therapeutic treatment in the field.

The technical prints abound with his work, his inventions, his discoveries. The following is only a partial list of terms now adopted and published in text books and technical works:

- Tesla two-phase, three phase, multi-phase, poly-phase system of power transmission

---

[12] John Pierpont Morgan (1837–1913) was an American financier and banker who dominated corporate finance on Wall Street throughout the late 19th and early 20th centuries. He created General Electric in 1891 and the US Steel Corporation in 1901 and held controlling interests in myriad other American businesses.

[13] Relating to work carried out or done with the hands.

- Tesla principle
- Tesla rotating magnetic field
- Tesla rotating magnetic field transformer
- Tesla induction motor
- Tesla split-phase motor
- Tesla system of distribution
- Tesla rotary transformer
- Tesla system of transformation by condenser discharges
- Tesla coil
- Tesla oscillation transformer
- Tesla electrical oscillator
- Tesla high frequency machines
- Tesla dynamo-electric oscillator
- Tesla tube
- Tesla lamp
- Tesla high-potential methods
- Tesla inductor
- Tesla marvels
- Tesla impedance phenomena
- Tesla electro-therapy
- Tesla electrical massage
- Tesla currents
- Tesla transmission
- Teslaic experiments
- Tesla capacity
- Tesla arc light system
- Tesla third brush regulation
- Tesla devices
- Tesla sparks
- Tesla arrangements
- Tesla theory
- Tesla point

MY INVENTIONS & other essays

- Tesla Steam Turbine
- Tesla Gas Turbine
- Tesla Water Turbine
- Tesla Pump
- Tesla Compressor
- Tesla Igniter
- Tesla condensers
- Tesla electro-static field
- Tesla effects
- Tesla wireless system
- Tesla methods of wireless transmission
- Tesla magnifying transmitter
- Tesla telautomata
- Tesla insulation
- Tesla underground transmission, etc.

The other night the Editors of the *Experimenter* had the opportunity of passing an evening with Tesla. We talked about many things, so interesting, that I will reserve them for another article — but mostly, of course, the conversation centered about Tesla himself.

"Dr. Tesla," I said to him, "you are aware of our great admiration for you, which may or may not be important. But the great public knows little of your mark. Even many of those technically educated — excuse the frankness — think that you are either a dreamer or, worse yet, crazy. The fact is the world does not understand you because you live in the next century. Moses was a great man, but the Bible teaches us that he was "heavy of tongue" and could not make himself understood. His brother therefore always spoke in his stead, announcing to his hearers what Moses had to say. Why not let the *Experimenter* be your brother? Why not let us translate your work into a language that the man in the street can readily understand? We have the knowledge and the technical training to do your inventions justice by means of graphic illustration and wash drawings.[14] The public does not want patent drawings or patent language. It wants pictures and plain English. You are a great inventor, but your 21st Century training prevents you from making yourself understood to a 20th Century public. My plan is to run one of your inventions every month, in plain English fully illustrated. That means that it will probably take over two years to deal with all of your more important inventions. At the end of this period the articles can be published in book form, a thing that does not exist at present. The plan is two-fold. First, the world at large will at last understand the highly important work you have accomplished and will fully recognize you. Second, it will be of greatest benefit to Science, to whom your inventions will then not be the sealed book they are today."

Knowing that Tesla had in the past continuously refused similar offers of dozens of great publishers of this country as well as abroad, I was not at all sanguine[15] of my own plan. Great was my surprise therefore, that he not only gave his consent, but he actually agreed to prepare each article personally with the Editors' collaboration.

Dr. Tesla wants it expressly understood that he is undertaking this great work chiefly to educate the young generation. He felt that he could not possibly reach such a large electrically trained young manhood, save through the medium of *Electrical Experimenter*. With its circulation above 100,000, all enthusiastic experimenters, Tesla feels that his greatest mission in life, namely, to assist our rising generation, will come near fulfillment.

Nikola Tesla's articles will therefore run serially every month in the *Electrical Experimenter*. The articles will be entitled: "My Inventions" — by Nikola Tesla. Every article will be entirely original; each will be illustrated with our own new illustrations

---

[14] A picture or sketch made by laying on washes of watercolor, typically in monochrome, over a pen or pencil drawing.

[15] Optimistic.

and with such wash drawings as made this journal so successful. The first article will appear in our February number.

We wish to congratulate *Experimenter* readers for having obtained for them probably the greatest technical news feature of a generation. I caution you: *Expect much!*

Hugo Gernsback
Electrical Experimenter
January 1919

NIKOLA TESLA
The Man

The door opens and out steps a tall figure — over six feet high — gaunt but erect. It approaches slowly, stately. You become conscious at once that you are face to face with a personality of a high order. Nikola Tesla advances and shakes your hand with a powerful grip, surprising for a man over sixty. A winning smile from piercing light blue-gray eyes, set in extraordinarily deep sockets, fascinates you and makes you feel at once at home.

You are guided into an office immaculate in its orderliness. Not a speck of dust is to be seen. No papers litter the desk, everything just so. It reflects the man himself, immaculate in attire, orderly and precise in his every movement. Dressed in a dark frock coat, he is entirely devoid of all jewelry. No ring, stickpin, or even watch-chain can be seen.

Tesla speaks — a very high almost falsetto voice. He speaks quickly and very convincingly. It is the man's voice chiefly which fascinates you.

As he speaks you find it difficult to take your eyes off his own. Only when he speaks to others do you have a chance to study his head, predominant of which is a very high forehead with a bulge between the eyes — the never-failing sign of an exceptional intelligence. Then the long, well-shaped nose, proclaiming the scientist.

How does this man, who has accomplished such a tremendous work, keep young and manage to surprise the world with more and more new inventions as he grows older? How does this youth of sixty, who is a professor of mathematics, a great mechanical and electrical engineer and the greatest inventor of all times, keep his physical as well as remarkable mental freshness?

To begin with, Tesla, who is by birth a Serbian, comes from a long-lived hardy race. His family tree abounds with centenarians.[1] Accordingly, Tesla — barring accidents — fully expects to be still inventing in A. D. 1960.

But the chief reason for his perpetual youth is found in his gastronomical frugality. Tesla has learned the great fundamental truth that most people not only eat all of their bodily ills, but actually eat themselves to death by either eating too much or else by food that does not agree with them.[2]

When Tesla found out that tobacco and black coffee interfered with his physical well-being, he quit both. This is the simple daily menu of the great inventor:

Breakfast: One to two pints of warm milk and a few eggs, prepared by himself — yes, he is a bachelor!

Lunch: None whatsoever, as a rule.

Dinner: Celery or the like, soup, a single piece of meat or fowl, potatoes and one other vegetable; a glass of light wine. For dessert, perhaps a slice of cheese, and invariably a big raw apple. And that's all.

Tesla is very fussy and particular about his food: he eats very little, but what he does eat must be of the very best. And he knows, for outside of being a great inventor in science he is an accomplished cook who has invented all sorts of savory dishes.

---

[1] Persons who are one hundred or more years old.
[2] If you research only one thing because of this book, make it this.

His only vice is his generosity. The man who, by the ignorant onlooker has often been called an idle dreamer, has made over a million dollars out of his inventions — and spent them as quickly on new ones. But Tesla is an idealist of the highest order and to such men money itself means but little.

Hugo Gernsback
Electrical Experimenter
February 1919

MY INVENTIONS

How does the world's greatest inventor invent? How does he carry out an invention? What sort of mentality has Nikola Tesla? Was his early life as commonplace as most of ours? What was the early training of one of the World's Chosen? These, and many other very interesting questions are answered in an incomparable manner by Nikola Tesla himself in this, his first article.

In his autobiography, treating mainly on his early youth, we obtain a good insight into the wonderful life this man has led. It reads like a fairy tale, which has the advantage of being true. For Tesla is no common mortal. He has led a charmed life — struck down by the pest, the cholera and what not — given up by doctors at least three times as dead — we find him at sixty, younger than ever. But — read his own words. You have never read the like before.

—Editor.
Electrical Experimenter
February 1919

I
My Early Life

The progressive development of man is vitally dependent on invention. It is the most important product of his creative brain. Its ultimate purpose is the complete mastery of mind over the material world, the harnessing of the forces of nature to human needs. This is the difficult task of the inventor who is often misunderstood and unrewarded. But he finds ample compensation in the pleasing exercises of his powers and in the knowledge of being one of that exceptionally privileged class without whom the race would have long ago perished in the bitter struggle against pitiless elements.

Speaking for myself, I have already had more than my full measure of this exquisite enjoyment, so much that for many years my life was little short of continuous rapture. I am credited with being one of the hardest workers and perhaps I am, if thought is the equivalent of labor, for I have devoted to it almost all of my waking hours. But if work is interpreted to be a definite performance in a specified time according to a rigid rule, then I may be the worst of idlers. Every effort under compulsion demands a sacrifice of life-energy. I never paid such a price. On the contrary, I have thrived on my thoughts.

In attempting to give a connected and faithful account of my activities in this series of articles which will be presented with the

MY INVENTIONS & other essays

assistance of the Editors of the *Electrical Experimenter* and are chiefly addressed to our young men readers, I must dwell, however reluctantly, on the impressions of my youth and the circumstances and events which have been instrumental in determining my career.

Our first endeavors are purely instinctive, promptings of an imagination vivid and undisciplined. As we grow older reason asserts itself and we become more and more systematic and designing. But those early impulses, though not immediately productive, are of the greatest moment and may shape our very destinies. Indeed, I feel now that had I understood and cultivated instead of suppressing them, I would have added substantial value to my bequest to the world. But not until I had attained manhood did I realize that I was an inventor.

This was due to a number of causes. In the first place I had a brother who was gifted to an extraordinary degree — one of those rare phenomena of mentality which biological investigation has failed to explain. His premature death left my parents disconsolate.[1] We owned a horse which had been presented to us by a dear friend. It was a magnificent animal of Arabian breed, possessed of almost human intelligence, and was cared for and petted by the whole family, having on one occasion saved my father's life under remarkable circumstances. My father had been called one winter night to perform an urgent duty and while crossing the mountains, infested by wolves, the horse became frightened and ran away, throwing him violently to the ground. It arrived home bleeding and exhausted, but after the alarm was sounded immediately dashed off again, returning to the spot, and before the searching party were far on the way they were met by my father, who had recovered consciousness and remounted, not realizing that he had been lying in the snow for several hours. This horse was responsible for my brother's injuries from which he died. I witnessed the tragic scene

---

[1] Without consolation or comfort; unhappy.

My Early Life

and although fifty-six years have elapsed since, my visual impression of it has lost none of its force. The recollection of his attainments made every effort of mine seem dull in comparison.

Anything I did that was creditable[2] merely caused my parents to feel their loss more keenly. So I grew up with little confidence in myself. But I was far from being considered a stupid boy, if I am to judge from an incident of which I have still a strong remembrance. One day the Aldermen[3] were passing through a street where I was at play with other boys. The oldest of these venerable gentlemen — a wealthy citizen — paused to give a silver piece to each of us. Coming to me he suddenly stopped and commanded, "Look in my eyes." I met his gaze, my hand outstretched to receive the much valued coin, when, to my dismay, he said, "No, not much, you can get nothing from me, you are too smart." They used to tell a funny story about me. I had two old aunts with wrinkled faces, one of them having two teeth protruding like the tusks of an elephant which she buried in my cheek every time she kissed me. Nothing would scare me more than the prospect of being hugged by these as affectionate as unattractive relatives. It happened that while being carried in my mother's arms they asked me who was the prettier of the two. After examining their faces intently, I answered thoughtfully, pointing to one of them, "This here is not as ugly as the other."

Then again, I was intended from my very birth for the clerical profession and this thought constantly oppressed me. I longed to be an engineer but my father was inflexible. He was the son of an officer who served in the army of the Great Napoleon and, in common with his brother, professor of mathematics in a prominent institution, had received a military education but, singularly enough, later embraced the clergy in which vocation he achieved eminence. He was a very erudite man, a veritable natural philosopher, poet

---

[2] Deserving public acknowledgment and praise but not necessarily outstanding or successful.

[3] Members of several municipal legislative bodies in a city or town.

ch. ends p. 13

and writer and his sermons were said to be as eloquent as those of Abraham a Sancta-Clara.[4] He had a prodigious memory and frequently recited at length from works in several languages. He often remarked playfully that if some of the classics were lost he could restore them. His style of writing was much admired. He penned sentences short and terse and was full of wit and satire. The humorous remarks he made were always peculiar and characteristic. Just to illustrate, I may mention one or two instances. Among the help there was a cross-eyed man called Mane, employed to do work around the farm. He was chopping wood one day. As he swung the axe, my father, who stood nearby and felt very uncomfortable, cautioned him, "For God's sake, Mane, do not strike at what you are looking but at what you intend to hit." On another occasion he was taking out for a drive a friend who carelessly permitted his costly fur coat to rub on the carriage wheel. My father reminded him of it saying, "Pull in your coat, you are ruining my tire." He had the odd habit of talking to himself and would often carry on an animated conversation and indulge in heated argument, changing the tone of his voice. A casual listener might have sworn that several people were in the room.

Although I must trace to my mother's influence whatever inventiveness I possess, the training he gave me must have been helpful. It comprised all sorts of exercises — as, guessing one another's thoughts, discovering the defects of some form or expression, repeating long sentences or performing mental calculations. These daily lessons were intended to strengthen memory and reason and especially to develop the critical sense, and were undoubtedly very beneficial.

My mother descended from one of the oldest families in the

---

[4] Abraham a Sancta Clara (born Johann Ulrich Megerle; 1644–1709), Augustinian friar, preacher, author of popular books of devotion, and "the harlequin of the pulpit," gained an early reputation for pulpit eloquence, speaking as a man of the people, presenting religious truths, even the most bitter, with such graphic charm that listeners of all social classes found pleasure in his sermons.

My Early Life

country and a line of inventors. Both her father and grandfather originated numerous implements for household, agricultural and other uses. She was a truly great woman, of rare skill, courage, and fortitude, who had braved the storms of life and passed through many a trying experience. When she was sixteen a virulent pestilence swept the country. Her father was called away to administer the last sacraments to the dying and during his absence she went alone to the assistance of a neighboring family who were stricken by the dread disease. All of the members, five in number, succumbed in rapid succession. She bathed, clothed and laid out the bodies, decorating them with flowers according to the custom of the country and when her father returned he found everything ready for a Christian burial. My mother was an inventor of the first order and would, I believe, have achieved great things had she not been so remote from modern life and its multifold[5] opportunities. She invented and constructed all kinds of tools and devices and wove the finest designs from thread which was spun by her. She even planted the seeds, raised the plants and separated the fibers herself. She worked indefatigably, from break of day till late at night, and most of the wearing apparel and furnishings of the home was the product of her hands. When she was past sixty, her fingers were still nimble enough to tie three knots in an eyelash.[6]

There was another and still more important reason for my late awakening. In my boyhood I suffered from a peculiar affliction due to the appearance of images, often accompanied by strong flashes of light, which marred the sight of real objects and interfered with my thought and action. They were pictures of things and scenes which I had really seen, never of those I imagined. When a word was spoken to me the image of the object it designated would present itself vividly to my vision and sometimes I was quite unable to

---
[5] Many and various.
[6] Three!

distinguish whether what I saw was tangible or not. This caused me great discomfort and anxiety. None of the students of psychology or physiology whom I have consulted could ever explain satisfactorily these phenomena. They seem to have been unique although I was probably predisposed as I know that my brother experienced a similar trouble. The theory I have formulated is that the images were the result of a reflex action from the brain on the retina under great excitation. They certainly were not hallucinations such as are produced in diseased and anguished minds, for in other respects I was normal and composed. To give an idea of my distress, suppose that I had witnessed a funeral or some such nerve-racking spectacle. Then, inevitably, in the stillness of night, a vivid picture of the scene would thrust itself before my eyes and persist despite all my efforts to banish it. Sometimes it would even remain fixed in space though I pushed my hand through it. If my explanation is correct, it should be able to project on a screen the image of any object one conceives and make it visible. Such an advance would revolutionize all human relations. I am convinced that this wonder can and will be accomplished in time to come; I may add that I have devoted much thought to the solution of the problem.[7]

To free myself of these tormenting appearances, I tried to concentrate my mind on something else I had seen, and in this way I would often obtain temporary relief; but in order to get it I had to

---

[7] As Tesla has explained it here, it would seem that he is suggesting the holographic projection of human thought. In "The Thought Recorder" on p. 12 of the May 1919 issue of *EE*, he elaborated: "Although I am clinging to ideals, my conception of the universe is, I fear, grossly materialistic. As stated in some of my published articles, I have satisfied myself thoroughly through careful observation carried on for many years that we are simply automata acting in obedience to external influences, without power or initiative. The brain is not an accumulator as commonly held in philosophy, and contains no records whatever of a phonographic or photographic kind. In other words, there is no stored knowledge or memory as usually conceived, our brains are blanks. The brain has merely the quality to respond, becoming more and more susceptible as the impressions are often repeated, this resulting in memory. ⋯▸

My Early Life

conjure continuously new images. It was not long before I found that I had exhausted all of those at my command; my "reel" had run out, as it were, because I had seen little of the world — only objects in my home and the immediate surroundings. As I performed these mental operations for the second or third time, in order to chase the appearances from my vision, the remedy gradually lost all its force. Then I instinctively commenced to make excursions beyond the limits of the small world of which I had knowledge, and I saw new scenes. These were at first very blurred and indistinct, and would flit away when I tried to concentrate my attention upon them, but by and by I succeeded in fixing them; they gained in strength and distinctness and finally assumed the concreteness of real things. I soon discovered that my best comfort was attained if I simply went on in my vision farther and farther, getting new impressions all the time, and so I began to travel — of course, in my mind. Every night (and sometimes during the day), when alone, I would start on my journeys — see new places, cities and countries — live there, meet people and make friendships and acquaintances and, however unbelievable, it is a fact that they were just as dear to me as those in actual life and not a bit less intense in their manifestations.

This I did constantly until I was about seventeen when my thoughts turned seriously to invention. Then I observed to my delight that I could visualize with the greatest facility. I needed no

"There is a possibility, however, which I have indicated years ago, that we may finally succeed in not only reading thoughts accurately, but reproducing faithfully every mental image. It can be done through the analysis of the retina, which is instrumental in conveying impressions to the nerve centers and, in my opinion, is also capable of serving as an indicator of the mental processes taking place within. Evidently, when an object is seen, consciousness of the external form can only be due to the fact that those cones and rods of the retina which are covered by the image are affected differently from the rest and it is a speculation not too hazardous to assume that visualization is accompanied by a reflex action on the retina which might be detected by suitable instruments. In this way it might also be possible to project the reflex image on a screen, and with further refinement, resorting to the principle involved in moving pictures, the continuous play of thoughts might be rendered visible, recorded, and at-will reproduced."

models, drawings, or experiments. I could picture them all as real in my mind. Thus I have been led unconsciously to evolve what I consider a new method of materializing inventive concepts and ideas, which is radically opposite to the purely experimental and is in my opinion ever so much more expeditious and efficient. The moment one constructs a device to carry into practice a crude idea he finds himself unavoidably engrossed with the details and defects of the apparatus. As he goes on improving and reconstructing, his force of concentration diminishes and he loses sight of the great underlying principle. Results may be obtained but always at the sacrifice of quality.

My method is different. I do not rush into actual work. When I get an idea I start at once building it up in my imagination. I change the construction, make improvements and operate the device in my mind. It is absolutely immaterial to me whether I run my turbine in thought or test it in my shop. I even note if it is out of balance. There is no difference whatever, the results are the same. In this way I am able to rapidly develop and perfect a conception without touching anything. When I have gone so far as to embody in the invention every possible improvement I can think of and see no fault anywhere, I put into concrete form this final product of my brain. Invariably my device works as I conceived that it should, and the experiment comes out exactly as I planned it. In twenty years there has not been a single exception. Why should it be otherwise? Engineering, electrical and mechanical, is positive in results. There is scarcely a subject that cannot be mathematically treated and the effects calculated or the results determined beforehand from the available theoretical and practical data. The carrying out into practice of a crude idea as is being generally done is, I hold, nothing but a waste of energy, money, and time.

My early affliction had, however, another compensation. The incessant mental exertion developed my powers of observation and

My Early Life 9

enabled me to discover a truth of great importance. I had noted that the appearance of images was always preceded by actual vision of scenes under peculiar and generally very exceptional conditions and I was impelled on each occasion to locate the original impulse. After a while this effort grew to be almost automatic and I gained great facility in connecting cause and effect. Soon I became aware, to my surprise, that every thought I conceived was suggested by an external impression. Not only this but all my actions were prompted in a similar way. In the course of time it became perfectly evident to me that I was merely an automaton endowed with power of movement, responding to the stimuli of the sense organs and thinking and acting accordingly. The practical result of this was the art of telautomatics which has been so far carried out only in an imperfect manner. Its latent possibilities will, however, be eventually shown. I have been since years planning self-controlled automata and believe that mechanisms can be produced which will act as if possessed of reason, to a limited degree, and will create a revolution in many commercial and industrial departments.

    I was about twelve years old when I first succeeded in banishing an image from my vision by willful effort, but I never had any control over the flashes of light to which I have referred. They were, perhaps, my strangest experience and inexplicable. They usually occurred when I found myself in a dangerous or distressing situation, or when I was greatly exhilarated. In some instances I have seen all the air around me filled with tongues of living flame. Their intensity, instead of diminishing, increased with time and seemingly attained a maximum when I was about twenty-five years old. While in Paris, in 1883, a prominent French manufacturer sent me an invitation to a shooting expedition which I accepted. I had been long confined to the factory and the fresh air had a wonderfully invigorating effect on me. On my return to the city that night I felt a positive sensation that my brain had caught fire. I saw a light as though a small

ch. ends p. 13

sun was located in it and I passed the whole night applying cold compressions to my tortured head. Finally the flashes diminished in frequency and force but it took more than three weeks before they wholly subsided. When a second invitation was extended to me my answer was an emphatic NO!

These luminous phenomena still manifest themselves from time to time, as when a new idea opening up possibilities strikes me, but they are no longer exciting, being of relatively small intensity. When I close my eyes I invariably observe first, a background of very dark and uniform blue, not unlike the sky on a clear but starless night. In a few seconds this field becomes animated with innumerable scintillating[8] flakes of green, arranged in several layers and advancing toward me. Then there appears, to the right, a beautiful pattern of two systems of parallel and closely spaced lines, at right angles to one another, in all sorts of colors with yellow-green and gold predominating. Immediately thereafter the lines grow brighter and the whole is thickly sprinkled with dots of twinkling light. This picture moves slowly across the field of vision and in about ten seconds vanishes to the left, leaving behind a ground of rather unpleasant and inert gray which quickly gives way to a billowy sea of clouds, seemingly trying to mold themselves in living shapes. It is curious that I cannot project a form into this gray until the second phase is reached. Every time, before falling asleep, images of persons or objects flit before my view. When I see them I know that I am about to lose consciousness. If they are absent and refuse to come it means a sleepless night.

To what an extent imagination played a part in my early life I may illustrate by another odd experience. Like most children I was fond of jumping and developed an intense desire to support myself in the air. Occasionally a strong wind richly charged with oxygen blew from the mountains rendering my body as light as

[8] In this context, sparkling or twinkling.

cork and then I would leap and float in space for a long time. It was a delightful sensation and my disappointment was keen when later I undeceived myself.

During that period I contracted many strange likes, dislikes, and habits, some of which I can trace to external impressions while others are unaccountable. I had a violent aversion against the earrings of women but other ornaments, as bracelets, pleased me more or less according to design. The sight of a pearl would almost give me a fit but I was fascinated with the glitter of crystals or objects with sharp edges and plane surfaces. I would not touch the hair of other people except, perhaps, at the point of a revolver. I would get a fever by looking at a peach and if a piece of camphor was anywhere in the house it caused me the keenest discomfort. Even now I am not insensible to some of these upsetting impulses. When I drop little squares of paper in a dish filled with liquid, I always sense a peculiar and awful taste in my mouth. I counted the steps in my walks and calculated the cubical contents of soup plates, coffee cups, and pieces of food — otherwise my meal was unenjoyable. All repeated acts or operations I performed had to be divisible by three and if I missed I felt impelled to do it all over again, even if it took hours.

Up to the age of eight years, my character was weak and vacillating.[9] I had neither courage or strength to form a firm resolve. My feelings came in waves and surges and vibrated unceasingly between extremes. My wishes were of consuming force and like the heads of the Hydra, they multiplied.[10] I was oppressed by thoughts of pain in life and death and religious fear. I was swayed by superstitious belief and lived in constant dread of the spirit of evil, of ghosts and ogres and other unholy monsters of the dark. Then, all at once, there came a tremendous change which altered the course of my whole

---
[9] Alternate or waver between different opinions or actions; be indecisive.
[10] In Greek and Roman Mythology, the Hydra was a many-headed snake with regenerative powers: for every head chopped off, it would regrow two.

existence. Of all things I liked books the best.[11] My father had a large library and whenever I could manage I tried to satisfy my passion for reading. He did not permit it and would fly into a rage when he caught me in the act. He hid the candles when he found that I was reading in secret. He did not want me to spoil my eyes. But I obtained tallow, made the wicking and cast the sticks into tin forms, and every night I would bush the keyhole and the cracks and read, often till dawn, when all others slept and my mother started on her arduous daily task. On one occasion I came across a novel entitled *Abafi* (the Son of Aba), a Serbian translation of a well known Hungarian writer, Josika.[12] This work somehow awakened my dormant powers of will and I began to practice self-control. At first my resolutions faded like snow in April, but in a little while I conquered my weakness and felt a pleasure I never knew before — that of doing as I willed. In the course of time this vigorous mental exercise became second nature. At the outset my wishes had to be subdued but gradually desire and will grew to be identical. After years of such discipline I gained so complete a mastery over myself that I toyed with passions which have meant destruction to some of the strongest men. At a certain age I contracted a mania for gambling which greatly worried my parents. To sit down to a game of cards was for me the quintessence of pleasure. My father led an exemplary life and could not excuse the senseless waste of time and money in which I indulged. I had a strong resolve but my philosophy was bad. I would say to him, "I can stop whenever I please but is it worthwhile to give up that which I would purchase with the joys of Paradise?" On frequent occasions he gave vent to his anger and contempt but my mother was different. She understood

---

[11] Agreed!

[12] Miklós Jósika (1794–1865) was a Hungarian soldier, politician, and author who cemented his literary reputation with his 1836 novel *Abafi*, a vivid depiction of Transylvania in the time of Sigismund Báthory (1573–1613), Prince of Transylvania, whose unpopular anti-Turkish policy led to civil war.

the character of men and knew that one's salvation could only be brought about through his own efforts. One afternoon, I remember, when I had lost all my money and was craving for a game, she came to me with a roll of bills and said, "Go and enjoy yourself. The sooner you lose all we possess the better it will be. I know that you will get over it." She was right. I conquered my passion then and there and only regretted that it had not been a hundred times as strong. I not only vanquished but tore it from my heart so as not to leave even a trace of desire. Ever since that time I have been as indifferent to any form of gambling as to picking teeth.

During another period I smoked excessively, threatening to ruin my health. Then my will asserted itself and I not only stopped but destroyed all inclination. Long ago I suffered from heart trouble until I discovered that it was due to the innocent cup of coffee I consumed every morning. I discontinued at once, though I confess it was not an easy task. In this way I checked and bridled other habits and passions and have not only preserved my life but derived an immense amount of satisfaction from what most men would consider privation and sacrifice.

After finishing the studies at the Polytechnic Institute and University[13] I had a complete nervous breakdown and while the malady lasted I observed many phenomena strange and unbelievable.

---

[13] Graz University of Technology, founded in 1811, is one of five universities in Styria, Austria. Tesla dropped out after the first semester of his third year and did not receive a degree.

Boys will be boys, the world over. The Boy Tesla was no exception to the universal rule, as this, his second autobiographical article clearly proves.

Mr. Tesla in his inimitable, delightful way, here paints with a literary artist's brush his own intimate boyhood in charming as well as vivid colors.

We have often heard of Tesla, the dreamer. But if he is entitled to the epithet, his early boyhood certainly fails to reveal it. Tesla did not allow much grass to grow under his feet while a boy, for he assuredly was a strenuous, red-blooded youngster.

You will wish to read all about the greatest inventor's early boyhood. It is doubly valuable because it comes from his own pen. We promise you an interesting twenty-minutes' entertainment.

—Editor.
Electrical Experimenter
March 1919

II
My First Efforts at Invention

I shall dwell briefly on these extraordinary experiences, on account of their possible interest to students of psychology and physiology and also because this period of agony was of the greatest consequence on my mental development and subsequent labors. But it is indispensable to first relate the circumstances and conditions which preceded them and in which might be found their partial explanation.

From childhood I was compelled to concentrate attention upon myself. This caused me much suffering but, to my present view, it was a blessing in disguise for it has taught me to appreciate the inestimable value of introspection in the preservation of life, as well as a means of achievement. The pressure of occupation and the incessant stream of impressions pouring into our consciousness through all the gateways of knowledge make modern existence hazardous in many ways. Most persons are so absorbed in the contemplation of the outside world that they are wholly oblivious to what is passing on within themselves.

The premature death of millions is primarily traceable to this cause. Even among those who exercise care it is a common mistake to avoid imaginary, and ignore the real dangers. And what is true of an individual also applies, more or less, to a people as a whole. Witness, in illustration, the Prohibition movement.[1] A drastic, if not unconstitutional, measure is now being put through in this country to prevent the consumption of alcohol and yet it is a positive fact that coffee, tea, tobacco, chewing gum and other stimulants, which are freely indulged in even at the tender age, are vastly more injurious to the national body, judging from the number of those who succumb. So, for instance, during my student years I gathered from the published necrologues[2] in Vienna, the home of coffee drinkers, that deaths from heart trouble sometimes reached *sixty-seven percent* of the total. Similar observations might probably be made in cities where the consumption of tea is excessive. These delicious beverages super-excite and gradually exhaust the fine fibers of the brain. They also interfere seriously with arterial circulation and should be enjoyed all the more sparingly as their deleterious[3] effects are slow and imperceptible. Tobacco, on the other hand, is conducive to easy and pleasant thinking and detracts from the intensity and concentration necessary to all original and vigorous effort of the intellect. Chewing gum is helpful for a short while but soon drains the glandular system and inflicts irreparable damage, not to speak of the revulsion it creates. Alcohol in small quantities is an excellent tonic, but is toxic in its action when absorbed in larger amounts, quite immaterial as to whether it is taken in as whiskey or produced in the stomach from sugar. But it should not be overlooked that all these are great eliminators assisting Nature, as they do, in upholding her stern but just law of the survival of the fittest. Eager reformers should also be mindful of the eternal

---

[1] The prevention by law of the manufacture and sale of alcohol, especially in the United States between 1920 and 1933.
[2] A list of obituaries or deaths.
[3] Harmful, injurious.

My First Efforts at Invention

perversity of mankind which makes the indifferent *"laissez-faire"*[4] by far preferable to enforced restraint.

The truth about this is that we need stimulants to do our best work under present living conditions, and that we must exercise moderation and control our appetites and inclinations in every direction. That is what I have been doing for many years, in this way maintaining myself young in body and mind. Abstinence was not always to my liking but I find ample reward in the agreeable experiences I am now making. Just in the hope of converting some to my precepts and convictions I will recall one or two.

A short time ago I was returning to my hotel. It was a bitter cold night, the ground slippery, and no taxi to be had. Half a block behind me followed another man, evidently as anxious as myself to get under cover. Suddenly my legs went up in the air. In the same instant there was a flash in my brain, the nerves responded, the muscles contracted, I swung through 180 degrees and landed on my hands. I resumed my walk as though nothing had happened, then the stranger caught up with me. "How old are you?" he asked, surveying me critically.

"Oh, about fifty-nine," I replied. "What of it?"

"Well," said he, "I have seen a cat do this but never a man."

About a month since I wanted to order new eyeglasses and went to an oculist[5] who put me through the usual tests. He looked at me incredulously as I read off with ease the smallest print at considerable distance. But when I told him that I was past sixty he gasped in astonishment. Friends of mine often remark that my suits fit me like gloves but they do not know that all my clothing is made to measurements which were taken nearly 35 years ago and never changed. During this same period my weight has not varied one pound.

---
[4] Translated from French, literally: allow to do. A policy or attitude of letting things take their own course, without interfering.

[5] Optometrist; eye doctor.

In this connection I may tell a funny story. One evening, in the winter of 1885, Mr. Edison, Edward H. Johnson, the President of the Edison Illuminating Company, Mr. Batchellor, Manager of the works, and myself entered a little place opposite 65 Fifth Avenue where the offices of the company were located. Someone suggested guessing weights and I was induced to step on a scale. Edison felt me all over and said: "Tesla weighs 152 lbs. to an ounce," and he guessed it exactly. Stripped, I weighed 142 lbs. and that is still my weight. I whispered to Mr. Johnson: "How is it possible that Edison could guess my weight so closely?"

"Well," he said, lowering his voice. "I will tell you, confidentially, but you must not say anything. He was employed for a long time in a Chicago slaughterhouse where he weighed thousands of hogs every day! That's why." My friend, the Hon. Chauncey M. Depew, tells of an Englishman on whom he sprung one of his original anecdotes and who listened with a puzzled expression but — a year later — laughed out loud. I will frankly confess it took me longer than that to appreciate Johnson's joke.

Now, my well being is simply the result of a careful and measured mode of living and perhaps the most astonishing thing is that three times in my youth I was rendered by illness a hopeless physical wreck and given up by physicians. More than this, through ignorance and lightheartedness, I got into all sorts of difficulties, dangers, and scrapes from which I extricated myself as by enchantment. I was almost drowned a dozen times; was nearly boiled alive and just missed being cremated. I was entombed, lost and frozen. I had hair-breadth escapes from mad dogs, hogs, and other wild animals. I passed through dreadful diseases and met with all kinds of odd mishaps and that I am hale and hearty today seems like a miracle. But as I recall these incidents to my mind I feel convinced that my preservation was not altogether accidental.

An inventor's endeavor is essentially lifesaving. Whether he

My First Efforts at Invention

harnesses forces, improves devices, or provides new comforts and conveniences, he is adding to the safety of our existence. He is also better qualified than the average individual to protect himself in peril, for he is observant and resourceful. If I had no other evidence that I was, in a measure, possessed of such qualities I would find it in these personal experiences. The reader will be able to judge for himself if I mention one or two instances. On one occasion, when about 14 years old, I wanted to scare some friends who were bathing with me. My plan was to dive under a long floating structure and slip out quietly at the other end. Swimming and diving came to me as naturally as to a duck and I was confident that I could perform the feat. Accordingly I plunged into the water and, when out of view, turned around and proceeded rapidly toward the opposite side. Thinking that I was safely beyond the structure, I rose to the surface but to my dismay struck a beam. Of course, I quickly dived and forged ahead with rapid strokes until my breath was beginning to give out. Rising for the second time, my head came again in contact with a beam. Now I was becoming desperate. However, summoning all my energy, I made a third frantic attempt but the result was the same. The torture of suppressed breathing was getting unendurable, my brain was reeling and I felt myself sinking. At that moment, when my situation seemed absolutely hopeless, I experienced one of those flashes of light and the structure above me appeared before my vision. I either discerned or guessed that there was a little space between the surface of the water and the boards resting on the beams and, with consciousness nearly gone, I floated up, pressed my mouth close to the planks and managed to inhale a little air, unfortunately mingled with a spray of water which nearly choked me. Several times I repeated this procedure as in a dream until my heart, which was racing at a terrible rate, quieted down and I gained composure. After that I made a number of unsuccessful dives, having completely lost the sense of direction,

ch. ends p. 27

but finally succeeded in getting out of the trap when my friends had already given me up and were fishing for my body.

That bathing season was spoiled for me through recklessness but I soon forgot the lesson and only two years later I fell into a worse predicament. There was a large flour mill with a dam across the river near the city where I was studying at that time. As a rule the height of the water was only two or three inches above the dam and to swim out to it was a sport not very dangerous in which I often indulged. One day I went alone to the river to enjoy myself as usual. When I was a short distance from the masonry, however, I was horrified to observe that the water had risen and was carrying me along swiftly. I tried to get away but it was too late. Luckily, though, I saved myself from being swept over by taking hold of the wall with both hands. The pressure against my chest was great and I was barely able to keep my head above the surface. Not a soul was in sight and my voice was lost in the roar of the fall. Slowly and gradually I became exhausted and unable to withstand the strain longer. Just as I was about to let go, to be dashed against the rocks below, I saw in a flash of light a familiar diagram illustrating the hydraulic principle that the pressure of a fluid in motion is proportionate to the area exposed, and automatically I turned on my left side. As if by magic the pressure was reduced and I found it comparatively easy in that position to resist the force of the stream. But the danger still confronted me. I knew that sooner or later I would be carried down, as it was not possible for any help to reach me in time, even if I attracted attention. I am ambidextrous now but then I was left-handed and had comparatively little strength in my right arm. For this reason I did not dare to turn on the other side to rest and nothing remained but to slowly push my body along the dam. I had to get away from the mill toward which my face was turned as the current there was much swifter and deeper. It was a long and painful ordeal and I came near to failing at its very end for I was confronted with a depression in the masonry. I managed to get over with the last ounce of my force and fell in a swoon when I reached the bank, where I was found. I had torn virtually all the skin from my left side and it took several weeks before the fever subsided and I was well. These are only two of many instances but they may be sufficient to show that had it not been for the inventor's instinct I would not have lived to tell this tale.

Interested people have often asked me how and when I began to invent. This I can only answer from my present recollection in the light of which the first attempt I recall was rather ambitious for it involved the invention of an *apparatus* and a *method*. In the former I was anticipated but the latter was original. It happened in this way. One of my playmates had come into the possession of a hook and fishing-tackle which created quite an excitement in the village, and the next morning all started out to catch frogs. I was left alone and deserted owing to a quarrel with this boy. I had never seen a real hook and pictured it as something wonderful, endowed with peculiar qualities, and was despairing not to be one of the party. Urged by necessity, I somehow got hold of a piece of soft iron wire, hammered the end to a sharp point between two stones, bent it into shape, and fastened it to a strong string. I then cut a rod, gathered some bait, and went down to the brook where there were frogs in abundance. But I could not catch any and was almost discouraged when it occurred to me to dangle the empty hook in front of a frog sitting on a stump. At first he collapsed but by and by his eyes bulged out and became bloodshot, he swelled to twice his normal size and made a vicious snap at the hook.

Immediately I pulled him up. I tried the same thing again and again and the method proved infallible. When my comrades, who in spite of their fine outfit had caught nothing, came to me they were green with envy. For a long time I kept my secret and enjoyed the monopoly but finally yielded to the spirit of Christmas.

Every boy could then do the same and the following summer brought disaster to the frogs.

In my next attempt I seem to have acted under the first instinctive impulse which later dominated me — to harness the energies of nature to the service of man. I did this through the medium of May-bugs — or June-bugs as they are called in America — which were a veritable pest in that country and sometimes broke the branches of trees by the sheer weight of their bodies. The bushes were black with them. I would attach as many as four of them to a crosspiece, rotably arranged on a thin spindle, and transmit the motion of the same to a large disc and so derive considerable "power." These creatures were remarkably efficient, for once they were started they had no sense to stop and continued whirling for hours and hours and the hotter it was the harder they worked. All went well until a strange boy came to the place. He was the son of a retired officer in the Austrian Army. That urchin ate May-bugs alive and enjoyed them as though they were the finest blue-point oysters. That disgusting sight terminated my endeavors in this promising field and I have never since been able to touch a May-bug or any other insect for that matter.

After that, I believe, I undertook to take apart and assemble the clocks of my grandfather. In the former operation I was always successful but often failed in the latter. So it came that he brought my work to a sudden halt in a manner not too delicate and it took thirty years before I tackled another clockwork again. Shortly thereafter I went into the manufacture of a kind of pop-gun which comprised a hollow tube, a piston, and two plugs of hemp. When firing the gun, the piston was pressed against the stomach and the tube was pushed back quickly with both hands. The air between the plugs was compressed and raised to high temperature and one of them was expelled with a loud report. The art consisted in selecting a tube of the proper taper from the hollow stalks. I did very well

My First Efforts at Invention

with that gun but my activities interfered with the window panes in our house and met with painful discouragement. If I remember rightly, I then took to carving swords from pieces of furniture which I could conveniently obtain. At that time I was under the sway of the Serbian national poetry and full of admiration for the feats of the heroes. I used to spend hours in mowing down my enemies in the form of cornstalks which ruined the crops and netted me several spankings from my mother. Moreover these were not of the formal kind but the genuine article.

    I had all this and more behind me before I was six years old and had passed through one year of elementary school in the village of Smiljan where I was born.[6] At this juncture we moved to the little city of Gospic nearby. This change of residence was like a calamity to me. It almost broke my heart to part from our pigeons, chickens, and sheep, and our magnificent flock of geese which used to rise to the clouds in the morning and return from the feeding grounds at sundown in battle formation, so perfect that it would have put a squadron of the best aviators of the present day to shame. In our new house I was but a prisoner, watching the strange people I saw through the window blinds. My bashfulness was such that I would rather have faced a roaring lion than one of the city dudes who strolled about. But my hardest trial came on Sunday when I had to dress up and attend the service. There I met with an accident, the mere thought of which made my blood curdle like sour milk for years afterward. It was my second adventure in a church. Not long before I was entombed for a night in an old chapel on an inaccessible mountain which was visited only once a year. It was an awful experience, but this one was worse. There was a wealthy lady in town, a good but pompous woman, who used to come to the church gorgeously painted up and attired with an enormous train and attendants. One Sunday I had just finished

---

[6] Smiljan is a village in the mountainous region of Western Lika in Croatia.

ringing the bell in the belfry and rushed downstairs when this grand dame was sweeping out and I jumped on her train. It tore off with a ripping noise which sounded like a salvo of musketry fired by raw recruits. My father was livid with rage. He gave me a gentle slap on the cheek, the only corporal punishment he ever administered to me but I almost feel it now. The embarrassment and confusion that followed are indescribable. I was practically ostracized until something else happened which redeemed me in the estimation of the community.

An enterprising young merchant had organized a fire department. A new fire engine was purchased, uniforms provided, and the men drilled for service and parade. The engine was, in reality, a pump to be worked by sixteen men and was beautifully painted red and black. One afternoon the official trial was prepared for and the machine was transported to the river. The entire population turned out to witness the great spectacle. When all the speeches and ceremonies were concluded, the command was given to pump, but not a drop of water came from the nozzle. The professors and experts tried in vain to locate the trouble. The fizzle was complete when I arrived at the scene. My knowledge of the mechanism was nil and I knew next to nothing of air pressure, but instinctively I felt for the suction hose in the water and found that it had collapsed. When I waded in the river and opened it up the water rushed forth and not a few Sunday clothes were spoiled. Archimedes running naked through the streets of Syracuse and shouting Eureka at the top of his voice did not make a greater impression than myself.[7] I was carried on the shoulders and was the hero of the day.

---

[7] Charged with proving that a new crown made for King Hiero II of Syracuse was not pure gold as the goldsmith had claimed, Archimedes (287–212 BC) is said to have invented a method for determining the volume of an object with an irregular shape as he stepped into a bathtub and realized the water displaced by his body was equal to the weight of his body. Forgetting that he was naked, he went running naked from his home to tell the king, shouting "Eureka!" all the way.

My First Efforts at Invention

Upon settling in the city I began a four-years' course in the so-called Normal School preparatory to my studies at the College or *Real Gymnasium*. During this period my boyish efforts and exploits, as well as troubles, continued. Among other things I attained the unique distinction of champion crow catcher in the country. My method of procedure was extremely simple. I would go in the forest, hide in the bushes, and imitate the call of the bird. Usually I would get several answers and in a short while a crow would flutter down into the shrubbery near me. After that all I needed to do was to throw a piece of cardboard to distract its attention, jump up and grab it before it could extricate itself from the undergrowth. In this way I would capture as many as I desired. But on one occasion something occurred which made me respect them. I had caught a fine pair of birds and was returning home with a friend.

This photograph shows in the background the house in which Mr. Tesla's family resided. The edifice at the right is the "Real Gymnasium" where he studied. The Ecclesiastic gentleman (white collar, center) is his uncle, the Metropolitan of Bosnia, who was a great statesman and who thwarted the designs of Austria upon Serbia at a critical period.

When we left the forest, thousands of crows had gathered making a frightful racket. In a few minutes they rose in pursuit and soon enveloped us. The fun lasted until all of a sudden I received a blow on the back of my head which knocked me down. Then they attacked me viciously. I was compelled to release the two birds and was glad to join my friend who had taken refuge in a cave.

In the schoolroom there were a few mechanical models which interested me and turned my attention to water turbines. I constructed many of these and found great pleasure in operating them. How extraordinary was my life an incident may illustrate. My uncle had no use for this kind of pastime and more than once rebuked me. I was fascinated by a description of Niagara Falls I had perused, and pictured in my imagination a big wheel run by the Falls. I told my uncle that I would go to America and carry out this scheme. Thirty years later I saw my ideas carried out at Niagara and marveled at the unfathomable mystery of the mind.

I made all kinds of other contrivances and contraptions but among these the arbalests[8] I produced were the best. My arrows, when shot, disappeared from sight and at close range traversed a plank of pine one inch thick. Through the continuous tightening of the bows I developed skin on my stomach very much like that of a crocodile and I am often wondering whether it is due to this exercise that I am able even now to digest cobblestones! Nor can I pass in silence my performances with the sling[9] which would have enabled me to give a stunning exhibit at the Hippodrome.[10] And now I will tell of one of my feats with this antique implement of war which will strain to the utmost the credulity of the reader. I was practicing while walking with my uncle along the river.

---

[8] Crossbows with a special mechanism for drawing back and releasing the string.
[9] Slingshot; a strap or loop, used to hurl stones or other small objects.
[10] The hippodrome was an ancient Greek stadium for horse racing and chariot racing.

The sun was setting, the trout were playful and from time to time one would shoot up into the air, its glistening body sharply defined against a projecting rock beyond. Of course any boy might have hit a fish under these propitious[11] conditions but I undertook a much more difficult task and I foretold to my uncle, to the minutest detail, what I intended doing: I was to hurl a stone to meet the fish, press its body against the rock, and cut it in two. It was no sooner said than done. My uncle looked at me almost scared out of his wits and exclaimed *"vade retro Satanas!"*[12] and it was a few days before he spoke to me again. Other records, however great, will be eclipsed but I feel that I could peacefully rest on my laurels for a thousand years.

---

[11] Giving or indicating a good chance of success; favorable.
[12] Translated from Latin: Get thee behind me, Satan!

This installment, no doubt the most interesting of the three published so far, reveals many extraordinary occurrences and experiences in the world's great inventor's life — experiences such as do not fall to the lot of ordinary mortals. And Tesla, the many sided, aside of inventing, knows the rare art of painting word-pictures. He does so here in a masterly fashion. He tells us how he finally conceived the induction motor — perhaps his greatest discovery — the invention which changed the face of the globe, the invention which made possible the street car, the subway, the electric train, power transmission, the harnessing of water falls, and countless others. But let Tesla tell you himself how it all came about. It is a classic worth reading.

—Editor.
Electrical Experimenter
April 1919

III
My Later Endeavors
The Discovery of the Rotating Magnetic Field

At the age of ten I entered the Real Gymnasium which was a new and fairly well equipped institution. In the department of physics were various models of classical scientific apparatus, electrical and mechanical. The demonstrations and experiments performed from time to time by the instructors fascinated me and were undoubtedly a powerful incentive to invention. I was also passionately fond of mathematical studies and often won the professor's praise for rapid calculation. This was due to my acquired facility of visualizing the figures and performing the operations, not in the usual intuitive manner, but as in actual life. Up to a certain degree of complexity it was absolutely the same to me whether I wrote the symbols on the board or conjured them before my mental vision. But freehand drawing, to which many hours of the course were devoted, was an annoyance I could not endure. This was rather remarkable as most of the members of the family excelled in it. Perhaps my aversion was simply due to the predilection I found in undisturbed thought. Had it not been for a few exceptionally stupid boys, who could not do anything at all, my record would have been the worst.

MY INVENTIONS & other essays

It was a serious handicap as under the then existing educational regime, drawing being obligatory, this deficiency threatened to spoil my whole career and my father had considerable trouble in railroading me from one class to another.

In the second year at that institution I became obsessed with the idea of producing continuous motion through steady air pressure. The pump incident, of which I have told, had set afire my youthful imagination and impressed me with the boundless abilities of a vacuum. I grew frantic in my desire to harness this inexhaustible energy but for a long time I was groping in the dark. Finally, however, my endeavors crystallized in an invention which was to enable me to achieve what no other mortal ever attempted.

Imagine a cylinder freely rotatable on two bearings and partly surrounded by a rectangular trough which fits it perfectly. The open side of the trough is closed by a partition so that the cylindrical segment within the enclosure divides the latter into two compartments entirely separated from each other by air-tight sliding joints. One of these compartments being sealed and once for all exhausted, the other remaining open, a perpetual rotation of the cylinder would result, at least, I thought so. A wooden model was constructed and fitted with infinite care and when I applied the pump on one side and actually observed that there was a tendency to turning, I was delirious with joy. Mechanical flight was the one thing I wanted to accomplish although still under the discouraging recollection of a bad fall I sustained by jumping with an umbrella from the top of a building. Every day I used to transport myself through the air to distant regions but could not understand just how I managed to do it. Now I had something concrete — a flying machine with nothing more than a rotating shaft, flapping wings, and — a vacuum of unlimited power! From that time on I made my daily aerial excursions in

My Later Endeavors

a vehicle of comfort and luxury as might have befitted King Solomon. It took years before I understood that the atmospheric pressure acted at right angles to the surface of the cylinder and that the slight rotary effort I observed was due to a leak. Though this knowledge came gradually it gave me a painful shock.

I had hardly completed my course at the Real Gymnasium when I was prostrated with a dangerous illness or rather, a score of them, and my condition became so desperate that I was given up by physicians. During this period I was permitted to read constantly, obtaining books from the Public Library which had been neglected and entrusted to me for classification of the works and preparation of the catalogs. One day I was handed a few volumes of new literature unlike anything I had ever read before and so captivating as to make me utterly forget my hopeless state. They were the earlier works of Mark Twain and to them might have been due the miraculous recovery which followed. Twenty-five years later, when I met Mr. Clemens and we formed a friendship between us, I told him of the experience and was amazed to see that great man of laughter burst into tears.

My studies were continued at the higher Real Gymnasium in Carlstadt, Croatia,[1] where one of my aunts resided. She was a distinguished lady, the wife of a Colonel who was an old war-horse having participated in many battles. I never can forget the three years I passed at their home. No fortress in time of war was under a more rigid discipline. I was fed like a canary bird. All the meals were of the highest quality and deliciously prepared but short in quantity by a thousand percent. The slices of ham cut by my aunt were like tissue paper. When the Colonel would put something substantial on my plate she would snatch it away and say excitedly

---

[1] Originally known as Karlstadt, Karlovac is a city in central Croatia known as the city of parks and the town on four rivers.

ch. ends p. 40

to him: "Be careful, Niko is very delicate." I had a voracious appetite and suffered like Tantalus.[2] But I lived in an atmosphere of refinement and artistic taste quite unusual for those times and conditions. The land was low and marshy and malaria fever never left me while there despite the enormous amounts of quinine[3] I consumed. Occasionally the river would rise and drive an army of rats into the buildings, devouring everything even to the bundles of the fierce paprika. These pests were to me a welcome diversion. I thinned their ranks by all sorts of means, which won me the unenviable distinction of rat-catcher in the community. At last, however, my course was completed, the misery ended, and I obtained the certificate of maturity which brought me to the crossroads.

During all those years my parents never wavered in their resolve to make me embrace the clergy, the mere thought of which filled me with dread. I had become intensely interested in electricity under the stimulating influence of my Professor of Physics, who was an ingenious man and often demonstrated the principles by apparatus of his own invention. Among these I recall a device in the shape of a freely rotatable bulb, with tinfoil coatings, which was made to spin rapidly when connected to a static machine. It is impossible for me to convey an adequate idea of the intensity of feeling I experienced in witnessing his exhibitions of these mysterious phenomena. Every impression produced a thousand echoes in my mind. I wanted to know more of this wonderful force; I longed for experiment and investigation and resigned myself to the inevitable with aching heart.

Just as I was making ready for the long journey home I received word that my father wished me to go on a shooting expedition. It was a strange request as he had been always strenuously

---

[2] In Greek mythology, Tantalus was famous for his punishment in Tartarus, where he was made to stand in a pool of water beneath a fruit tree with low branches, with the fruit ever eluding his grasp, and the water always receding before he could take a drink.

[3] A bitter crystalline compound present in cinchona bark (South American evergreen tree or shrub), used as a tonic and formerly as an antimalarial drug.

opposed to this kind of sport. But a few days later I learned that the cholera was raging in that district and, taking advantage of an opportunity, I returned to Gospic in disregard of my parents' wishes. It is incredible how absolutely ignorant people were as to the causes of this scourge which visited the country in intervals of from fifteen to twenty years. They thought that the deadly agents were transmitted through the air and filled it with pungent odors and smoke. In the meantime they drank the infected water and died in heaps. I contracted the awful disease on the very day of my arrival and although surviving the crisis, I was confined to bed for nine months with scarcely any ability to move. My energy was completely exhausted and for the second time I found myself at death's door. In one of the sinking spells which was thought to be the last, my father rushed into the room. I still see his pallid face as he tried to cheer me in tones belying his assurance. "Perhaps," I said, "I may get well if you will let me study engineering."

"You will go to the best technical institution in the world," he solemnly replied, and I knew that he meant it. A heavy weight was lifted from my mind but the relief would have come too late had it not been for a marvelous cure brought about through a bitter decoction[4] of a peculiar bean. I came to life like another Lazarus to the utter amazement of everybody.

My father insisted that I spend a year in healthful physical outdoor exercises to which I reluctantly consented. For most of this term I roamed in the mountains, loaded with a hunter's outfit and a bundle of books, and this contact with nature made me stronger in body as well as in mind. I thought and planned, and conceived many ideas almost as a rule delusive. The vision was clear enough but the knowledge of principles was very limited. In one of my inventions I proposed to convey letters and packages across the seas, through a submarine tube, in spherical containers of sufficient

[4] The liquor resulting from concentrating the essence of a substance by heating or boiling, especially a medicinal preparation made from a plant.

strength to resist the hydraulic pressure. The pumping plant, intended to force the water through the tube, was accurately figured and designed and all other particulars carefully worked out. Only one trifling detail, of no consequence, was lightly dismissed. I assumed an arbitrary velocity of the water and, what is more, took pleasure in making it high, thus arriving at a stupendous performance supported by faultless calculations. Subsequent reflections, however, on the resistance of pipes to fluid flow determined me to make this invention public property.

Another one of my projects was to construct a ring around the equator which would, of course, float freely and could be arrested in its spinning motion by reactionary forces, thus enabling travel at a rate of about one thousand miles an hour, impracticable by rail. The reader will smile. The plan was difficult of execution, I will admit, but not nearly so bad as that of a well-known New York professor, who wanted to pump the air from the torrid to the temperate zones, entirely forgetful of the fact that the Lord had provided a gigantic machine for this very purpose.

Still another scheme, far more important and attractive, was to derive power from the rotational energy of terrestrial bodies. I had discovered that objects on the earth's surface, owing to the diurnal[5] rotation of the globe, are carried by the same alternately in and against the direction of translatory movement.[6] From this results a great change in momentum which could be utilized in the simplest imaginable manner to furnish motive effort in any habitable region of the world. I cannot find words to describe my disappointment when later I realized that I was in the predicament of Archimedes, who vainly sought for a fixed point in the universe.

At the termination of my vacation I was sent to the Polytechnic School in Gratz, Styria, which my father had chosen as one of the

---

[5] Daily.
[6] In physics, the motion of a body in which every point of the body moves parallel to and the same distance as every other point of the body.

My Later Endeavors   35

oldest and best reputed institutions. That was the moment I had eagerly awaited and I began my studies under good auspices and firmly resolved to succeed. My previous training was above the average, due to my father's teaching and opportunities afforded. I had acquired the knowledge of a number of languages and waded through the books of several libraries, picking up information more or less useful. Then again, for the first time, I could choose my subjects as I liked, and freehand drawing was to bother me no more.

I had made up my mind to give my parents a surprise, and during the whole first year I regularly started my work at three o'clock in the morning and continued until eleven at night, no Sundays or holidays excepted. As most of my fellow-students took things easily, naturally enough I eclipsed all records. In the course of that year I past through nine exams and the professors thought I deserved more than the highest qualifications. Armed with their flattering certificates, I went home for a short rest, expecting a triumph, and was mortified when my father made light of these hard won honors. That almost killed my ambition; but later, after he had died, I was pained to find a package of letters which the professors had written him to the effect that unless he took me away from the Institution I would be killed through overwork.

Thereafter I devoted myself chiefly to physics, mechanics, and mathematical studies, spending the hours of leisure in the libraries. I had a veritable mania for finishing whatever I began, which often got me into difficulties. On one occasion I started to read the works of Voltaire when I learned, to my dismay, that there were close on one hundred large volumes in small print which that monster had written while drinking seventy-two cups of black coffee per diem.[7] It had to be done, but when I laid aside the last book I was very glad, and said, "Nevermore!"

My first year's showing had won me the appreciation and

---
[7] Voltaire: Beast Mode!

ch. ends p. 40

friendship of several professors. Among these were Prof. Rogner, who was teaching arithmetical subjects and geometry; Prof. Poeschl, who held the chair of theoretical and experimental physics, and Dr. Alle, who taught integral calculus and specialized in differential equations. This scientist was the most brilliant lecturer to whom I ever listened. He took a special interest in my progress and would frequently remain for an hour or two in the lecture room, giving me problems to solve, in which I delighted. To him I explained a flying machine I had conceived, not an illusionary invention, but one based on sound, scientific principles, which has become realizable through my turbine and will soon be given to the world. Both Professors Rogner and Poeschl were curious men. The former had peculiar ways of expressing himself and whenever he did so there was a riot, followed by a long and embarrassing pause. Prof. Poeschl was a methodical and thoroughly grounded German. He had enormous feet and hands like the paws of a bear, but all of his experiments were skillfully performed with lock-like precision and without a miss.

It was in the second year of my studies that we received a Gramme dynamo from Paris,[8] having the horseshoe form of a laminated field magnet, and a wire-wound armature with a commutator. It was connected up and various effects of the currents were shown. While Prof. Poeschl was making demonstrations, running the machine as a motor, the brushes gave trouble, sparking badly, and I observed that it might be possible to operate a motor without these appliances. But he declared that it could not be done and did me the honor of delivering a lecture on the subject, at the conclusion of which he remarked: "Mr. Tesla may accomplish great things, but he certainly never will do this. It would be equivalent to converting a steadily

---

[8] Also known as a Gramme machine, Gramme ring, or Gramme magneto; an electrical generator that produces direct current, named for its Belgian inventor, Zénobe Gramme (1826–1901). It was the first generator to produce power on a commercial scale for industry.

My Later Endeavors

pulling force, like that of gravity, into a rotary effort. It is a perpetual motion scheme, an impossible idea." But instinct is something which transcends knowledge. We have, undoubtedly, certain finer fibers that enable us to perceive truths when logical deduction, or any other willful effort of the brain, is futile. For a time I wavered, impressed by the professor's authority, but soon became convinced I was right and undertook the task with all the fire and boundless confidence of youth.

I started by first picturing in my mind a direct-current machine, running it and following the changing flow of the currents in the armature. Then I would imagine an alternator and investigate the processes taking place in a similar manner. Next I would visualize systems comprising motors and generators and operate them in various ways. The images I saw were to me perfectly real and tangible. All my remaining term in Gratz was passed in intense but fruitless efforts of this kind, and I almost came to the conclusion that the problem was insolvable.

In 1880 I went to Prague, Bohemia, carrying out my father's wish to complete my education at the University there. It was in that city that I made a decided advance, which consisted in detaching the commutator from the machine and studying the phenomena in this new aspect, but still without result. In the year following there was a sudden change in my views of life. I realized that my parents had been making too great sacrifices on my account and resolved to relieve them of the burden. The wave of the American telephone had just reached the European continent and the system was to be installed in Budapest, Hungary. It appeared an ideal opportunity, all the more as a friend of our family was at the head of the enterprise. It was here that I suffered the complete breakdown of the nerves to which I have referred.

What I experienced during the period of that illness surpasses all belief. My sight and hearing were always extraordinary. I could

ch. ends p. 40

clearly discern objects in the distance when others saw no trace of them. Several times in my boyhood I saved the houses of our neighbors from fire by hearing the faint crackling sounds which did not disturb their sleep, and calling for help.

In 1899, when I was past forty and carrying on my experiments in Colorado, I could hear very distinctly thunderclaps at a distance of 550 miles. The limit of audition for my young assistants was scarcely more than 150 miles. My ear was thus over thirteen times more sensitive. Yet at that time I was, so to speak, stone deaf in comparison with the acuteness of my hearing while under the nervous strain. In Budapest I could hear the ticking of a watch with three rooms between me and the timepiece. A fly alighting on a table in the room would cause a dull thud in my ear. A carriage passing at a distance of a few miles fairly shook my whole body. The whistle of a locomotive twenty or thirty miles away made the bench or chair on which I sat vibrate so strongly that the pain was unbearable. The ground under my feet trembled continuously. I had to support my bed on rubber cushions to get any rest at all. The roaring noises from near and far often produced the effect of spoken words which would have frightened me had I not been able to resolve them into their accidental components. The sun's rays, when periodically intercepted, would cause blows of such force on my brain that they would stun me. I had to summon all my will power to pass under a bridge or other structure as I experienced a crushing pressure on the skull. In the dark I had the sense of a bat and could detect the presence of an object at a distance of twelve feet by a peculiar creepy sensation on the forehead. My pulse varied from a few to two hundred and sixty beats and all the tissues of the body quivered with twitchings and tremors which was perhaps the hardest to bear. A renowned physician who gave me daily large doses of Bromide of Potassium[9] pronounced my malady unique and incurable.

---

[9] Potassium bromide (KBr) is a salt, widely used as an anticonvulsant and a sedative in the late 19th and early 20th centuries.

My Later Endeavors

It is my eternal regret that I was not under the observation of experts in physiology and psychology at that time. I clung desperately to life, but never expected to recover. Can anyone believe that so hopeless a physical wreck could ever be transformed into a man of astonishing strength and tenacity, able to work thirty-eight years almost without a day's interruption, and find himself still strong and fresh in body and mind? Such is my case. A powerful desire to live and to continue the work, and the assistance of a devoted friend and athlete accomplished the wonder. My health returned and with it the vigor of mind. In attacking the problem again I almost regretted that the struggle was soon to end. I had so much energy to spare. When I undertook the task it was not with a resolve such as men often make. With me it was a sacred vow, a question of life and death. I knew that I would perish if I failed. Now I felt that the battle was won. Back in the deep recesses of the brain was the solution, but I could not yet give it outward expression. One afternoon, which is ever present in my recollection, I was enjoying a walk with my friend in the City Park and reciting poetry. At that age I knew entire books by heart, word for word. One of these was Goethe's *Faust*.[10] The sun was just setting and reminded me of the glorious passage:

Sie rückt und weicht, der Tag ist überlebt,
Dort eilt sie hin und fördert neues Leben.
O! daß kein Flügel mich vom Boden hebt
Ihr nach und immer nach zu streben![11]

---

[10] A tragic play in two parts by Johann Wolfgang von Goethe (1749–1832) that tells the story of a man who sold his soul to the Devil in exchange for earthly fulfillment. Considered by many to be Goethe's magnum opus and the greatest work of German literature.
[11] Translated from German: The glow retreats, done is the day of toil; It yonder hastes, new fields of life exploring; Ah, that no wing can lift me from the soil Upon its track to follow, follow soaring!

ch. ends next p.

*   *   *

*Ein schöner Traum indessen sie entweicht.
Ach! zu des Geistes Flügeln wird so leicht
Kein körperlicher Flügel sich gesellen!*[12]

As I uttered these inspiring words the idea came like a flash of lightning and in an instant the truth was revealed. I drew with a stick on the sand the diagrams shown six years later in my address before the American Institute of Electrical Engineers, and my companion understood them perfectly. The images I saw were wonderfully sharp and clear and had the solidity of metal and stone, so much so that I told him: "See my motor here; watch me reverse it." I cannot begin to describe my emotions. Pygmalion seeing his statue come to life could not have been more deeply moved.[13] A thousand secrets of nature which I might have stumbled upon accidentally I would have given for that one which I had wrested from her against all odds and at the peril of my existence.

---

[12] Translated from German: A glorious dream! though now the glories fade. Alas! the wings that lift the mind no aid; Of wings to lift the body can bequeath me.

[13] In Greek mythology, and most familiar from Ovid's narrative poem *Metamorphoses*, Pygmalion was a king and a sculptor who fell in love with a female statue he carved from ivory.

Nikola Tesla — Age 23

Nikola Tesla — Age 29

Nikola Tesla — Age 39

The proverbial trials and tribulations known to every inventor were not spared Tesla, the world's greatest inventor of all times. In this article we see him, arrived at young manhood, struggling along in the cold world. Already his fame has spread far and wide and his genius is recognized. But converting genius and fame into dollars and cents is quite a different matter, and the world is full of unappreciative and unscrupulous men. Tesla, the idealist, cared little for money and thus was promptly taken advantage of. But let Tesla himself tell you in his own inimitable style. It is a wonderful story.

In this month's installment Tesla also tells us how he made one of his most important as well as sensational discoveries — the Tesla Coil. Few inventions have caused such a sensation as this one which culminated in the only man-made lightning ever produced. The Tesla coil has so many uses and has been built in so many styles that it would take a catalog to list them all. For the spectacular high frequency stunts on the stage down to the "violet" ray machine in your home: all are Tesla coils in one form or another.

Wireless without the Tesla Coil would not be possible today. Without an oscillation transformer, spark gap, and condenser — which is a Tesla Coil — the sending station would be crippled.

But it is for industrial purposes where the Tesla Coil will shine brightest in the future. The production of Ozone, the extraction of Nitrogen from the air in huge quantities — all are children of Tesla's fertile brain. His coil is the key to them all.

—Editor.
Electrical Experimenter
May 1919

IV
The Discovery of the Tesla Coil and Transformer

For a while I gave myself up entirely to the intense enjoyment of picturing machines and devising new forms. It was a mental state of happiness about as complete as I have ever known in life. Ideas came in an uninterrupted stream and the only difficulty I had was to hold them fast. The pieces of apparatus I conceived were to me absolutely real and tangible in every detail, even to the minute marks and signs of wear. I delighted in imagining the motors constantly running, for in this way they presented to mind's eye a more fascinating sight. When natural inclination develops into a passionate desire, one advances toward his goal in seven-league boots.[1] In less than two months I evolved virtually all the types of motors and modifications of the system which are now identified with my name. It was, perhaps, providential that the necessities of existence commanded a temporary halt to this consuming activity of the mind. I came to Budapest prompted by a premature report concerning the telephone enterprise and, as irony of fate willed

---
[1] In European folklore, seven-league boots allow their wearer to take strides of seven leagues per step (a league was a unit of length roughly equal to the distance a person could walk in one hour).

it, I had to accept a position as draftsman in the Central Telegraph Office of the Hungarian Government at a salary which I deem it my privilege not to disclose! Fortunately, I soon won the interest of the Inspector-in-Chief and was thereafter employed on calculations, designs and estimates in connection with new installations, until the Telephone Exchange was started, when I took charge of the same. The knowledge and practical experience I gained in the course of this work was most valuable and the employment gave me ample opportunities for the exercise of my inventive faculties. I made several improvements in the Central Station apparatus and perfected a telephone repeater or amplifier which was never patented or publicly described but would be creditable to me even today. In recognition of my efficient assistance the organizer of the undertaking, Mr. Puskas, upon disposing of his business in Budapest, offered me a position in Paris which I gladly accepted.

I never can forget the deep impression that magic city produced on my mind. For several days after my arrival I roamed through the streets in utter bewilderment of the new spectacle. The attractions were many and irresistible, but, alas, the income was spent as soon as received. When Mr. Puskas asked me how I was getting along in the new sphere, I described the situation accurately in the statement that "the last twenty-nine days of the month are the toughest!" I led a rather strenuous life in what would now be termed "Rooseveltian fashion." Every morning, regardless of weather, I would go from the Boulevard St. Marcel, where I resided, to a bathing house on the Seine, plunge into the water, loop the circuit twenty-seven times and then walk an hour to reach Ivry, where the Company's factory was located. There I would have a woodchopper's breakfast at half-past seven o'clock and then eagerly await the lunch hour, in the meanwhile cracking hard nuts for the Manager of the Works, Mr. Charles Batchellor, who was an intimate friend and assistant of Edison. Here I was thrown in contact with a few Americans who fairly fell in love with me because of my proficiency in billiards. To these men I explained my invention and one of them, Mr. D. Cunningham, Foreman of the Mechanical Department, offered to form a stock company. The proposal seemed to me comical in the extreme. I did not have the faintest conception of what that meant except that it was an American way of doing things. Nothing came of it, however, and during the next few months I had to travel from one to another place in France and Germany to cure the ills of the power plants. On my return to Paris I submitted to one of the administrators of the Company, Mr. Rau, a plan for improving their dynamos and was given an opportunity. My success was complete and the delighted directors accorded me the privilege of developing automatic regulators which were much desired. Shortly after there was some trouble with the lighting plant which had been installed at the new railroad station in Strassburg, Alsace.[2] The wiring was defective and on the occasion of the opening ceremonies a large part of a wall was blown out through a short-circuit right in the presence of old Emperor William I.[3] The German Government refused to take the plant and the French Company was facing a serious loss. On account of my knowledge of the German language and past experience, I was entrusted with the difficult task of straightening out matters and early in 1883 I went to Strassburg on that mission.

Some of the incidents in that city have left an indelible record on my memory. By a curious coincidence, a number of men who subsequently achieved fame, lived there about that time. In later life I used to say, "There were bacteria of greatness in that old town. Others caught the disease but I escaped!" The practical work, correspondence, and conferences with officials kept me preoccupied day and night, but, as soon as I was able to manage I undertook the

[2] Strasbourg is located in Eastern France, at the border with Germany.
[3] William I or Wilhelm I (Wilhelm Friedrich Ludwig; 1797–1888) of the House of Hohenzollern was the first head of state of a united Germany and King of Prussia from 1861 until his death.

48 MY INVENTIONS & other essays

construction of a simple motor in a mechanical shop opposite the railroad station, having brought with me from Paris some material for that purpose. The consummation of the experiment was, however, delayed until the summer of that year when I finally had the satisfaction of *seeing rotation effected by alternating currents of different phase, and without sliding contacts or commutator,*

Tesla's first induction motor. This historic model is one of the two first presented before the American Institute of Electrical Engineers.

WHAT IS THE INDUCTION MOTOR?

The induction motor operates on alternating current. It has no commutator like a direct current motor, nor slip rings like an alternating current motor. Contrary to the two types just cited, the "field" current is not steady, but the current itself rotates constantly pulling around with it — by induction — the only moving part of the motor — the rotor or armature. Having no armature or slip rings, the induction motor never sparks. It consequently knows no "brush" trouble. It needs no attention because of its ruggedness. Only the bearings wear out. Its efficiency, too, is higher. On account of all this the induction motor is used in a preponderating proportion in street cars, electric trains, factories, etc.

as I had conceived a year before. It was an exquisite pleasure but not to compare with the delirium of joy following the first revelation.

Among my new friends was the former Mayor of the city, Mr. Bauzin, whom I had already in a measure acquainted with this and other inventions of mine and whose support I endeavored to enlist. He was sincerely devoted to me and put my project before several wealthy persons but, to my mortification, found no response. He wanted to help me in every possible way and the approach of the first of July, 1919, happens to remind me of a form of "assistance" I received from that charming man, which was not financial but nonetheless appreciated. In 1870, when the Germans invaded the country, Mr. Bauzin had buried a good sized allotment of St. Estephe of 1801[4] and he came to the conclusion that he knew no worthier person than myself to consume that precious beverage. This, I may say, is one of the unforgettable incidents to which I have referred. My friend urged me to return to Paris as soon as possible and seek support there. This I was anxious to do but my work and negotiations were protracted owing to all sorts of petty obstacles I encountered so that at times the situation seemed hopeless.

Just to give an idea of German thoroughness and "efficiency," I may mention here a rather funny experience. An incandescent lamp of 16 c.p.[5] was to be placed in a hallway and upon selecting the proper location I ordered the *monteur*[6] to run the wires. After working for a while he concluded that the engineer had to be consulted and this was done. The latter made several objections but ultimately agreed that the lamp should be placed two inches from the spot I had assigned, whereupon the work proceeded. Then the engineer became worried and told me that Inspector Averdeck should be notified. That important person called, investigated, debated, and

---
[4] Saint-Estèphe is a certified designation for red wine in the Bordeaux region of France.
[5] Candle Power: illuminating power expressed in candelas or candles.
[6] Translated from German: assembler; installer.

decided that the lamp should be shifted back two inches, which was the place I had marked. It was not long, however, before Averdeck got cold feet himself and advised me that he had informed *Ober-Inspector* Hieronimus of the matter and that I should await his decision. It was several days before the *Ober-Inspector* was able to free himself of other pressing duties but at last he arrived and a two-hour debate followed, when he decided to move the lamp two inches farther. My hopes that this was the final act were shattered when the *Ober-Inspector* returned and said to me: "*Regierungsrath* Funke is so particular that I would not dare to give an order for placing this lamp without his explicit approval." Accordingly arrangements for a visit from that great man were made. We started cleaning up and polishing early in the morning. Everybody brushed up, I put on my gloves and when Funke came with his retinue[7] he was ceremoniously received. After two hours' deliberation he suddenly exclaimed: "I must be going," and pointing to a place on the ceiling, he ordered me to put the lamp there. It was the exact spot which I had originally chosen.

So it went day after day with variations, but I was determined to achieve at whatever cost and in the end my efforts were rewarded. By the spring of 1884 all the differences were adjusted, the plant formally accepted, and I returned to Paris with pleasing anticipations. One of the administrators had promised me a liberal compensation in case I succeeded, as well as a fair consideration of the improvements I had made in their dynamos and I hoped to realize a substantial sum. There were three administrators whom I shall designate as A, B, and C for convenience. When I called on A *he* told me that B had the say. This gentleman thought that only C could decide and the latter was quite sure that A alone had the power to act. After several laps of this *circulus vitiosus*[8] it dawned upon me that my

---

[7] A group of advisers, assistants, or others accompanying an important person.
[8] Translated from Latin: vicious circle.

reward was a castle in Spain. The utter failure of my attempts to raise capital for development was another disappointment and when Mr. Batchellor pressed me to go to America with a view of redesigning the Edison machines, I determined to try my fortunes in the Land of Golden Promise. But the chance was nearly missed. I liquefied my modest assets, secured accommodations, and found myself at the railroad station as the train was pulling out. At that moment I discovered that my money and tickets were gone. What to do was the question. Hercules had plenty of time to deliberate but I had to decide while running alongside the train with opposite feelings surging in my brain like condenser oscillations. Resolve, helped by dexterity, won out in the nick of time and upon passing through the usual experiences, as trivial as unpleasant, I managed to embark for New York with the remnants of my belongings, some poems and articles I had written, and a package of calculations relating to solutions of an unsolvable integral and to my flying machine. During the voyage I sat most of the time at the stern of the ship watching for an opportunity to save somebody from a watery grave, without the slightest thought of danger. Later when I had absorbed some of the practical American sense I shivered at the recollection and marveled at my former folly.

I wish that I could put in words my first impressions of this country. In the Arabian Tales[9] I read how genii[10] transported people into a land of dreams to live through delightful adventures. My case was just the reverse. The genii had carried me from a world of dreams into one of realities. What I had left was beautiful, artistic and fascinating in every way; what I saw here was machined, rough and unattractive. A burly policeman was twirling his stick which looked to me as big as a log. I approached him politely with

[9] A collection of Middle Eastern folktales, also known as *One Thousand and One Nights*, compiled in Arabic during the Islamic Golden Age.
[10] Plural form of genie (pronounced gee–nee–eye).

the request to direct me. "Six blocks down, then to the left," he said, with murder in his eyes.

"Is this America?" I asked myself in painful surprise. "It is a century behind Europe in civilization." When I went abroad in 1889 — five years having elapsed since my arrival here — I became convinced that *it was more than one hundred years ahead of Europe* and nothing has happened to this day to change my opinion.

The meeting with Edison was a memorable event in my life. I was amazed at this wonderful man who, without early advantages and scientific training, had accomplished so much. I had studied a dozen languages, delved in literature and art, and had spent my best years in libraries reading all sorts of stuff that fell into my hands, from Newton's *Principia*[11] to the novels of Paul de Kock,[12] and felt that most of my life had been squandered. But it did not take long before I recognized that it was the best thing I could have done. Within a few weeks I had won Edison's confidence and it came about in this way.

The *S.S. Oregon*, the fastest passenger steamer at that time, had both of its lighting machines disabled and its sailing was delayed. As the superstructure had been built after their installation it was impossible to remove them from the hold. The predicament was a serious one and Edison was much annoyed. In the evening I took the necessary instruments with me and went aboard the vessel where I stayed for the night. The dynamos were in bad condition, having several short-circuits and breaks, but with the assistance of the crew I succeeded in putting them in good shape. At five o'clock

---

[11] *Philosophiae Naturalis Principia Mathematica* (Mathematical Principles of Natural Philosophy) by Isaac Newton (1642–1727), often referred to as simply the *Principia*, is a work in three books written in Latin, first published in 1687.

[12] Charles-Paul de Kock (1793–1871) was a prolific and extraordinarily popular French novelist, having written approximately 100 volumes, mostly concerning middle-class Parisian life. His most famous novels are *André le Savoyard* (1825) and *Le Barbier de Paris* (1826).

in the morning, when passing along Fifth Avenue on my way to the shop, I met Edison with Batchellor and a few others as they were returning home to retire. "Here is our Parisian running around at night," he said. When I told him that I was coming from the *Oregon* and had repaired both machines, he looked at me in silence and walked away without another word. But when he had gone some distance I heard him remark: "Batchellor, this is a damn good man," and from that time on I had full freedom in directing the work. For nearly a year my regular hours were from 10.30 A.M. until 5 o'clock the next morning without a day's exception. Edison said to me: "I have had many hard-working assistants but you take the cake." During this period I designed twenty-four different types of standard machines with short cores and of uniform pattern which replaced the old ones. The Manager had promised me fifty thousand dollars on the completion of this task but it turned out to be a practical joke. This gave me a painful shock and I resigned my position.

Immediately thereafter some people approached me with the proposal of forming an arc light company under my name, to which I agreed. Here finally was an opportunity to develop the motor, but when I broached the subject to my new associates they said: "No, we want the arc lamp. We don't care for this alternating current of yours." In 1886 my system of arc lighting was perfected and adopted for factory and municipal lighting, and I was free, but with no other possession than a beautifully engraved certificate of stock of hypothetical value. Then followed a period of struggle in the new medium for which I was not fitted, but the reward came in the end and in April, 1887, the Tesla Electric Company was organized, providing a laboratory and facilities. The motors I built there were exactly as I had imagined them. I made no attempt to improve the design, but merely reproduced the pictures as they appeared to my vision and the operation was always as I expected.

In the early part of 1888 an arrangement was made with the

54   MY INVENTIONS & other essays

Westinghouse Company for the manufacture of the motors on a large scale. But great difficulties had still to be overcome. My system was based on the use of low frequency currents and the Westinghouse experts had adopted 133 cycles with the object of securing advantages in the transformation. They did not want to depart from their standard forms of apparatus and my efforts had to be concentrated upon adapting the motor to these conditions. Another necessity was to produce a motor capable of running efficiently at this frequency on two wires which was not easy of accomplishment.

At the close of 1889, however, my services in Pittsburgh being no longer essential, I returned to New York and resumed experimental work in a laboratory on Grand Street, where I began immediately the design of high frequency machines. The problems of construction in this unexplored field were novel and quite peculiar and I encountered many difficulties. I rejected the inductor type, fearing that it might not yield perfect sine waves which were so important to resonant action. Had it not been for this I could have saved myself a great deal of labor. Another discouraging feature of the high frequency alternator seemed to be the inconstancy of speed which threatened to impose serious limitations to its use. I had already noted in my demonstrations before the American Institution of Electrical Engineers that several times the tune was lost, necessitating readjustment, and did not yet foresee, what I discovered long afterward, a means of operating a machine of this kind at a speed constant to such a degree as not to vary more than a small fraction of one revolution between the extremes of load.

From many other considerations it appeared desirable to invent a simpler device for the production of electric oscillations. In 1856 Lord Kelvin had exposed the theory of the condenser discharge, but no practical application of that important knowledge was made. I saw the possibilities and undertook the development of induction apparatus on this principle. My progress was so rapid as to enable me to exhibit at my lecture in 1891 a coil giving sparks of *five inches*. On that occasion I frankly told the engineers of a defect involved in the transformation by the new method, namely, the loss in the spark gap. Subsequent investigation showed that no matter what medium is employed, be it air, hydrogen, mercury vapor, oil, or a stream of electrons, the efficiency is the same. It is a law very much like that governing the conversion of mechanical energy. We may drop a weight from a certain height vertically down or carry it to the lower level along any devious path, it is immaterial insofar as the amount of work is concerned. Fortunately however, this drawback is not fatal as by proper proportioning of the resonant circuits an *efficiency of 85 percent* is attainable. Since my early announcement of the invention it has come into universal use and wrought a revolution in many departments. But a still greater future awaits it. When in 1900 I obtained powerful discharges of 100 feet and flashed a current around the globe, I was reminded of the first tiny spark I observed in my Grand Street laboratory and was thrilled by sensations akin to those I felt when I discovered the *rotating magnetic field*.

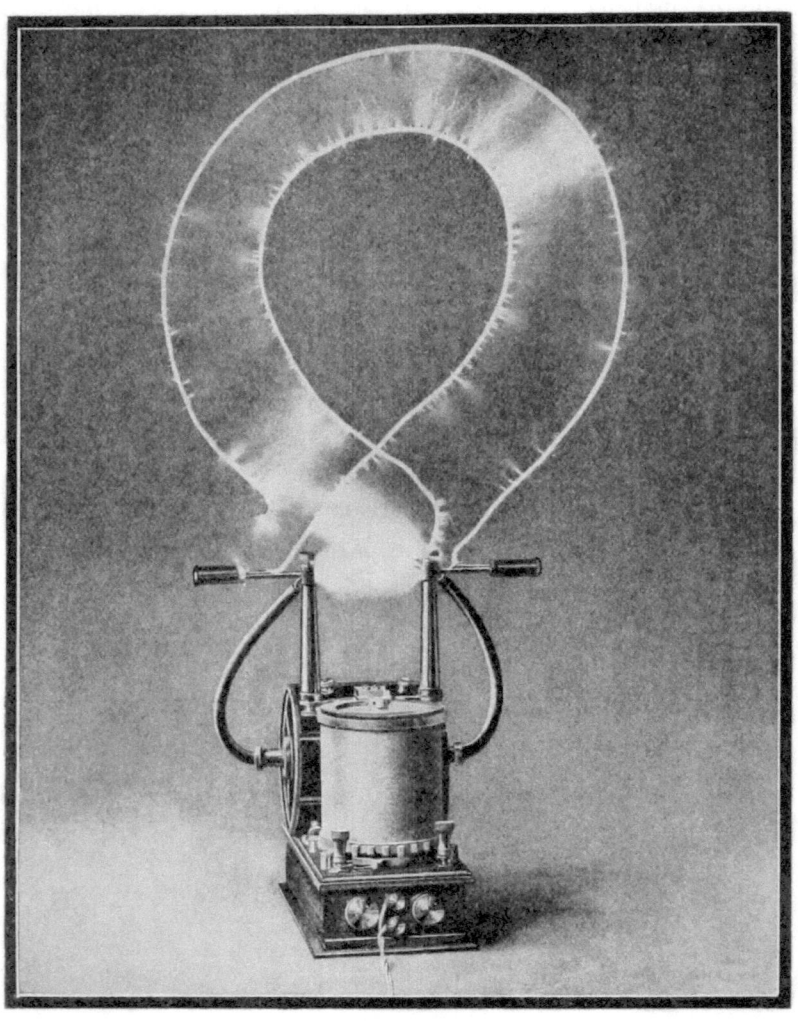

Tesla Oscillation Transformer (Tesla Coil) presented by Lord Kelvin before the British Association in August 1897. This small and compact instrument, only 8 inches high, developed two square feet of streamers with twenty-five watts from the 110 Volt DC supply circuit. The instrument contains a Tesla primary and secondary condenser and circuit controller.

This illustrates tests with spark discharges from a ball of 40cm radius in Tesla's Wireless Plant erected at Colorado Springs in 1899. The ball is connected to the free end of a grounded resonant circuit 17m in diameter. The disruptive potential of a ball in volts, according to Tesla, is approximately $V = 75,400\, r$ (r being in centimeters), that is, in this case 75,400 x 40 = 3,016,000 Volts. The gigantic Tesla Coil which produced these bolts of Thor was capable of furnishing a current of 1,100 Amperes in the high tension secondary. The primary coil had a diameter of 51 feet! This Tesla Coil produced discharges which were the nearest approach to lightning ever made by man.

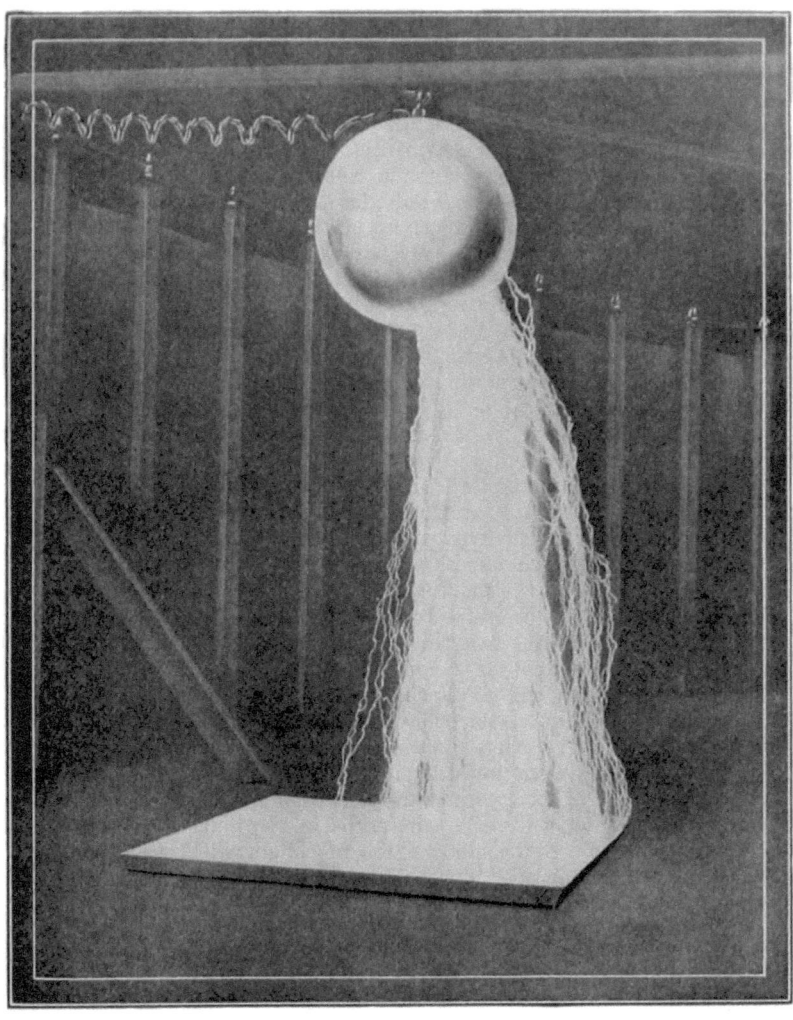

This revolutionary improvement was exhibited and explained by Tesla for the first time in his lecture before the American Institute of Electrical Engineers, May 20, 1891. It has made possible to generate automatically damped or undamped oscillations of any desired frequency and, what is equally important, of perfectly constant period. It has been instrumental in many great achievements and its use has become universal. The underlying principle may be briefly stated as follows: A source of electricity is made to charge a condenser and when the difference of potential at the terminals of the latter has reached a predetermined value, an air-gap is bridged, permitting the accumulated energy to be discharged through a circuit under resonant conditions, this resulting in a long series of isochronous impulses. These are either directly used or converted to any desired volume or pressure by means of a second circuit inductively linked with the first and tuned to the same. The diagram at right is taken from Tesla's lecture before the Franklin Institute and National Electric Light Association in 1893 and shows more elaborate arrangements of circuits, now quite familiar, for the conversion of ordinary direct or alternating currents into high frequency oscillations by this general method. In the mechanical apparatus illustrated, an attempt is made to convey an idea of the electrical operations as closely as practicable. The reciprocating and centrifugal pumps, respectively, represent an alternating and a direct current generator. The water takes the place of the electric fluid. The cylinder with its elastically restrained piston represents the condenser. The inertia of the moving parts corresponds to the self-induction of the electric circuit and the wide ports around the cylinder, through which the fluid can escape, perform the function of the air-gap. The operation of this apparatus will now be readily understood. Suppose first that the water is admitted to the cylinder from the centrifugal pump, this corresponding to the action of a continuous current generator. As the fluid is forced into the cylinder, the piston moves upward until the ports are uncovered, when a great quantity of the fluid rushes out, suddenly reducing the pressure so that the force of the compressed spring asserts itself and sends the piston down, closing the ports, whereupon these operations are repeated in as rapid succession as it may be desired. Each time the system, comprising the piston, rod, weights, and adjustable spring, receives a blow, it quivers at its own rate which is determined by the inertia of the moving parts and the pliability of the spring exactly as in the electrical system the period of the circuit is determined by the self-induction and capacity. Under the best conditions the natural period of the elastic system will be the same as that of the primarily impressed oscillations, and then the energy of the movement will be greatest. If, instead of the centrifugal, the reciprocating pump is employed, the operation is the same in principle except that the periodic impulses of the pump impose certain limitations. The best results are again obtained when synchronism is maintained between these and the natural oscillations of the system.

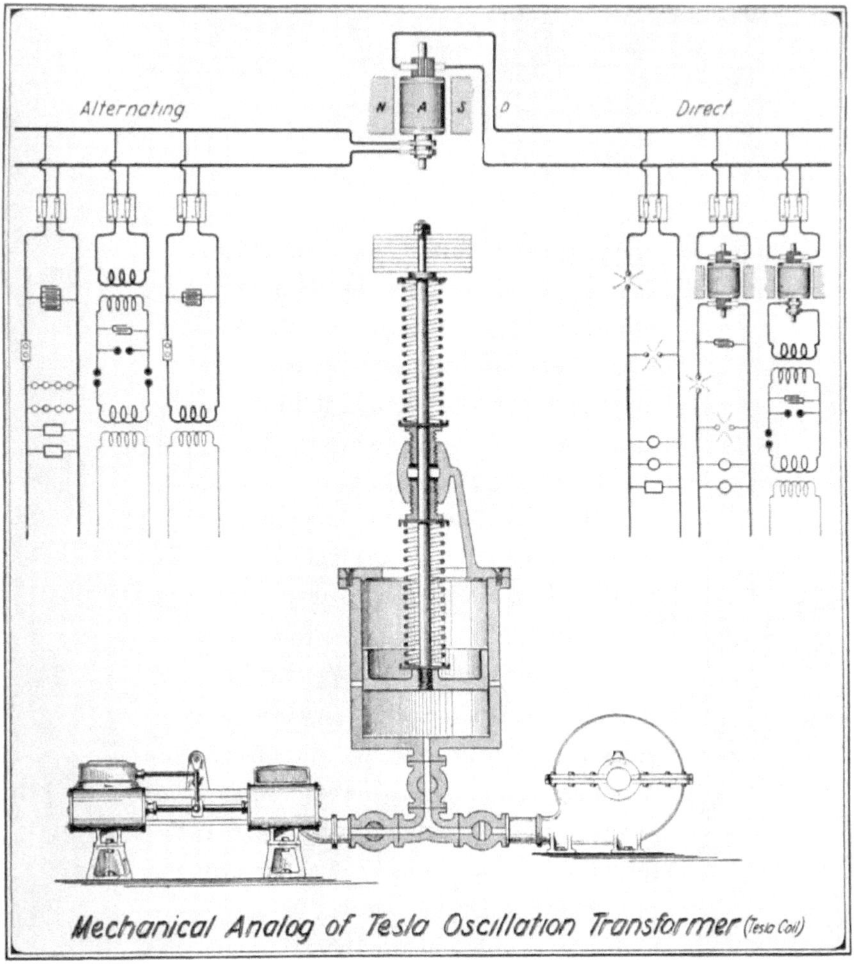

Alternating
N
A
S
D
Direct
Mechanical Analog of Tesla Oscillation Transformer (Tesla Coil)

contacts
primary of Tesla Coil
terminal of condenser
terminal of condenser
adjustable iron core
to direct or alternating current
condenser

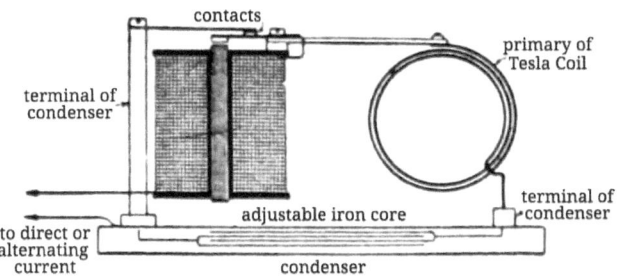

The purpose of this analog is to portray, as faithfully as possible, the phenomena of Tesla's Rotating Magnetic Field so as to make them easily understandable to the average reader. The two alternating fluxes are represented by streams of water having the same relation as to phase, amplitude, and direction. The magnetic polarity of the rotor is imitated by the employment of a body so shaped as to behave, with respect to the streams, exactly as the rotor with respect to the poles. Moreover, the corresponding rotating and stationary parts are given a similar appearance and are disposed in like manner. To make the analog complete, it may further be assumed that the liquid is compressible so that there will be a phase displacement between pressure and flow as that existing between electro-motive force and current.

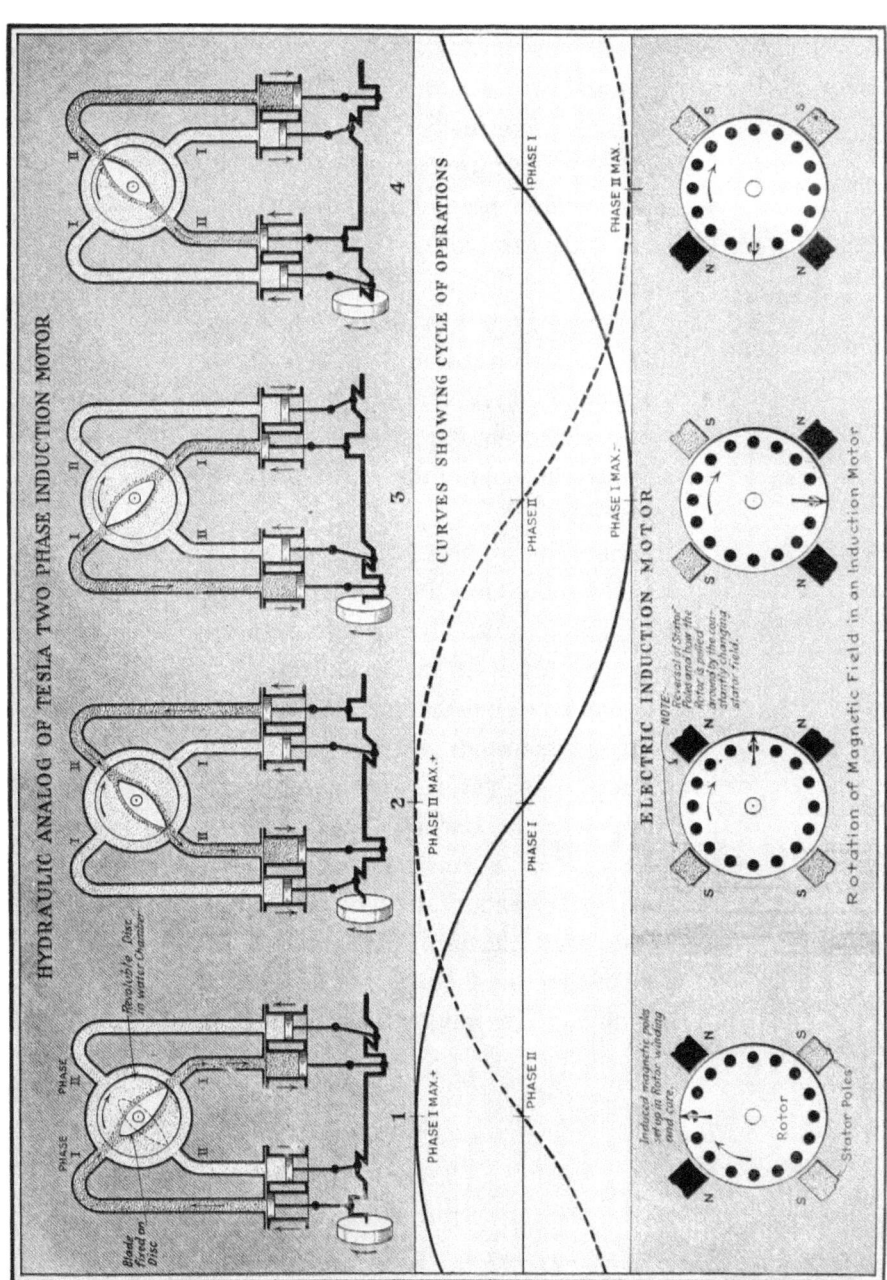

Imagine a man a century ago, bold enough to design and actually build a huge tower with which to transmit the human voice, music, pictures, press news, and even power through the earth to any distance whatever without wires! He probably would have been hung or burnt at the stake. So when Tesla built his famous tower on Long Island he was a hundred years ahead of his time. And foolish ridicule by our latter day arm-chair "savants" does not in the least mar Tesla's greatness.

    The titanic brain of Tesla has hardly produced a more amazing wonder than this "magnifying transmitter." Contrary to popular belief his tower was not built to radiate Hertzian waves into the ether. Tesla's system sends out thousands of horsepower though the earth — he has shown experimentally how power can be sent without wires over distances from a central point. Nor is there any mystery about it how he accomplishes the result. His historic U.S. patents and articles describe the method used. Tesla's Magnifying Transmitter is truly a modern lamp of Aladdin.

—Editor.
Electrical Experimenter
June 1919

V
The Magnifying Transmitter

As I review the events of my past life I realize how subtle are the influences that shape our destinies. An incident of my youth may serve to illustrate. One winter's day I managed to climb a steep mountain, in company with other boys. The snow was quite deep and a warm southerly wind made it just suitable for our purpose. We amused ourselves by throwing balls which would roll down a certain distance, gathering more or less snow, and we tried to outdo one another in this exciting sport. Suddenly a ball was seen to go beyond the limit, swelling to enormous proportions until it became as big as a house and plunged thundering into the valley below with a force that made the ground tremble. I looked on spellbound, incapable of understanding what had happened. For weeks afterward the picture of the avalanche was before my eyes and I wondered how anything so small could grow to such an immense size. Ever since that time the magnification of feeble actions fascinated me, and when, years later, I took up the experimental study of mechanical and electrical resonance, I was keenly interested from the very start. Possibly, had it not been for that early powerful impression,

I might not have followed up the little spark I obtained with my coil and never developed my best invention, the true history of which I'll tell here for the first time.

"Lionhunters" have often asked me which of my discoveries I prize most. This depends on the point of view. Not a few technical men, very able in their special departments, but dominated by a pedantic spirit and nearsighted, have asserted that excepting the induction motor I have given to the world little of practical use. This is a grievous mistake. A new idea must not be judged by its immediate results. My alternating system of power transmission came at a psychological moment, as a long-sought answer to pressing industrial questions, and although considerable resistance had to be overcome and opposing interests reconciled, as usual, the commercial introduction could not be long delayed. Now, compare this situation with that confronting my turbine, for example. One should think that so simple and beautiful an invention, possessing many features of an ideal motor, should be adopted at once and, undoubtedly, it would under similar conditions. But the prospective effect of the rotating field was not to render worthless existing machinery; on the contrary, it was to give it additional value. The system lent itself to new enterprise as well as to improvement of the old. My turbine is an advance of a character entirely different. It is a radical departure in the sense that its success would mean the abandonment of the antiquated types of prime movers on which billions of dollars have been spent. Under such circumstances the progress must be slow and perhaps the greatest impediment is encountered in the prejudicial opinions created in the minds of experts by organized opposition.

Only the other day I had a disheartening experience when I met my friend and former assistant, Charles F. Scott,[1] now professor of Electrical Engineering at Yale. I had not seen him for a long time and was glad to have an opportunity for a little chat at my office. Our conversation naturally enough drifted on my turbine and I became heated to a high degree. "Scott," I exclaimed, carried away by the vision of a glorious future, "my turbine will scrap all the heat-engines in the world." Scott stroked his chin and looked away thoughtfully, as though making a mental calculation. "That will make quite a pile of scrap," he said, and left without another word!

These and other inventions of mine, however, were nothing more than steps forward in certain directions. In evolving them I simply followed the inborn sense to improve the present devices without any special thought of our far more imperative necessities. The "Magnifying Transmitter" was the product of labors extending through years, having for their chief object the solution of problems which are infinitely more important to mankind than mere industrial development.

If my memory serves me right, it was in November 1890, that I performed a laboratory experiment which was one of the most extraordinary and spectacular ever recorded in the annals of science. In investigating the behavior of high frequency currents I had satisfied myself that an electric field of sufficient intensity could be produced in a room to light up electrodeless vacuum tubes. Accordingly, a transformer was built to test the theory and the first trial proved a marvelous success. It is difficult to appreciate what those strange phenomena meant at that time. We crave for new sensations but soon become indifferent to them. The wonders of yesterday

---

[1] Charles Felton Scott (1864–1944) was an electrical engineer, professor at Yale University, president of the American Institute of Electrical Engineers (AIEE), and known for his invention of the Scott-T transformer (a type of circuit used to produce two-phase electric power from a three-phase source, or vice versa).

are today common occurrences. When my tubes were first publicly exhibited they were viewed with amazement impossible to describe. From all parts of the world I received urgent invitations and numerous honors and other flattering inducements were offered to me, which I declined.

But in 1892 the demands became irresistible and I went to London where I delivered a lecture before the Institution of Electrical Engineers. It had been my intention to leave immediately for Paris in compliance with a similar obligation, but Sir James Dewar[2] insisted on my appearing before the Royal Institution.[3] I was a man of firm resolve but succumbed easily to the forceful arguments of the great Scotsman. He pushed me into a chair and poured out half a glass of a wonderful brown fluid which sparkled in all sorts of iridescent colors and tasted like nectar. "Now," said he, "you are sitting in Faraday's chair and you are enjoying whiskey he used to drink."[4] In both aspects it was an enviable experience. The next evening I gave a demonstration before that Institution, at the termination of which Lord Rayleigh[5] addressed the audience and his generous words gave me the first start in these endeavors. I fled from London and later from Paris to escape favors showered upon me, and journeyed to my home where I passed through a most painful ordeal and illness. Upon regaining my health I began to formulate

---

[2] Sir James Dewar (1842–1923) was a Scottish chemist and physicist, best known for his invention of the vacuum flask, which he used in conjunction with research into the liquefaction of gases.

[3] The Royal Society, formally The Royal Society of London for Improving Natural Knowledge, founded in 1660, is a learned society and the United Kingdom's national academy of sciences. It is the oldest national scientific institution in the world.

[4] Michael Faraday (refer to footnote #1 on p. xiv). As for what his favorite whiskey was, we Heathens would love to know!

[5] John William Strutt, 3rd Baron Rayleigh (1842–1919) was a British scientist who made extensive contributions to both theoretical and experimental physics and received the 1904 Nobel Prize in Physics. He spent his entire academic career at the University of Cambridge and served as President of the Royal Society from 1905 to 1908.

plans for the resumption of work in America. Up to that time I never realized that I possessed any particular gift of discovery but Lord Rayleigh, whom I always considered as an ideal man of science, had said so and if that was the case I felt that I should concentrate on some big idea.

One day, as I was roaming in the mountains, I sought shelter from an approaching storm. The sky became overhung with heavy clouds but somehow the rain was delayed until, all of a sudden, there was a lightning flash and, a few moments after, a deluge. This observation set me thinking. It was manifest that the two phenomena were closely related, as cause and effect, and a little reflection led me to the conclusion that the electrical energy involved in the precipitation of the water was inconsiderable, the function of lightning being much like that of a sensitive trigger.

Here was a stupendous possibility of achievement. If we could produce electric effects of the required quality, this whole planet and the conditions of existence on it could be transformed. The sun raises the water of the oceans and winds drive it to distant regions where it remains in a state of most delicate balance. If it were in our power to upset it when and wherever desired, this mighty life-sustaining stream could be at will controlled. We could irrigate arid deserts, create lakes and rivers, and provide motive power in unlimited amounts. This would be the most efficient way of harnessing the sun to the uses of man. The consummation depended on our ability to develop electric forces of the order of those in nature.[6] It seemed a hopeless undertaking, but I made up my mind to try it and immediately on my return to the United States, in the Summer of 1892, work was begun which was to me all the more attractive, because a means of the same kind was necessary for the successful transmission of energy without wires.

The first gratifying result was obtained in the spring of the

---

[6] In short: weather modification or controlling the weather.

succeeding year when I reached tensions of about 1,000,000 volts with my conical coil. That was not much in the light of the present art, but it was then considered a feat. Steady progress was made until the destruction of my laboratory by fire in 1895, as may be judged from an article by T. C. Martin[7] which appeared in the April number[8] of *The Century Magazine*.[9] This calamity set me back in many ways and most of that year had to be devoted to planning and reconstruction. However, as soon as circumstances permitted, I returned to the task.

Although I knew that higher electro-motive forces were attainable with apparatus of larger dimensions, I had an instinctive perception that the object could be accomplished by the proper design of a comparatively small and compact transformer. In carrying on tests with a *secondary in the form of a flat spiral*, as illustrated in my patents, the absence of streamers surprised me, and it was not long before I discovered that this was due to the position of the turns and their mutual action. Profiting from this observation I resorted to the use of a high tension conductor with turns of considerable diameter sufficiently separated to keep down the distributed capacity, while at the same time preventing undue accumulation of the charge at any point. The application of this principle enabled me to produce pressures of 4,000,000 volts, which was about the limit obtainable in my new laboratory at Houston Street, as the discharges extended through a distance of 16 feet. A photograph of this transmitter (at right) was published in the *Electrical Review* of November, 1898.[10]

[7] Thomas Commerford Martin (1856–1924) was an American electrical engineer, editor, and the first person to create a comprehensive compilation of Tesla's work with his 1894 book *The Inventions, Researches and Writings of Nikola Tesla*.
[8] Martin, T.C. (1895, April). Tesla's Oscillator and Other Inventions. *The Century Magazine*, 49(6), 916–933.
[9] *The Century Magazine* (1881–1930) was a New York-based, highly successful monthly magazine noted for its articles, illustrations, and fiction.
[10] Tesla's System of Electric Power Transmission Through Natural Media. (1898, November 11). *The Electrical Review*, 43(1094), 709-711.

View of model transformer, or "oscillator," photographed in action. Actual width of space traversed by the luminous streams issuing from a circular single terminal, over 16 feet. Area covered by the streamers, approximately 200 square feet. Estimated effective electrical pressure, 2½ million volts.

Rear view of Tesla's Colorado Springs Experimental Station.

In order to advance further along this line I had to go into the open, and in the spring of 1899, having completed preparations for the erection of a wireless plant, I went to Colorado where I remained for more than one year. Here I introduced other improvements and refinements which made it possible to generate currents of any tension that may be desired. Those who are interested will find some information in regard to the experiments I conducted there in my article, "The Problem of Increasing Human Energy" in *The Century Magazine* of June 1900, to which I have referred on a previous occasion.[11]

I have been asked by the *Electrical Experimenter* to be quite explicit on this subject so that my young friends among the readers of the magazine will clearly understand the construction and operation of my "Magnifying Transmitter" and the purposes for which it is intended. Well, then, in the first place, it is a *resonant transformer* with a secondary in which the parts, charged to a high potential, are of considerable area and arranged in space along ideal enveloping surfaces of very large radii of curvature, and at proper distances from one another thereby insuring a *small electric surface density everywhere* so that *no leak can occur even if the conductor is bare*. It is suitable for any frequency, from a few to many thousands of cycles per second, and can be used in the production of currents of tremendous volume and moderate pressure, or of smaller amperage and immense electromotive force. The maximum electric *tension is merely dependent on the curvature of the surfaces* on which the charged elements are situated and the area of the latter.

Judging from my past experience, as much as 100,000,000 volts are perfectly practicable. On the other hand currents of many thousands of amperes may be obtained in the antenna. A plant of but very moderate dimensions is required for such performances. Theoretically, a terminal of less than 90 feet in diameter is sufficient to

---

[11] Tesla, N. (1900, June). The Problem of Increasing Human Energy. *The Century Magazine*, 60(2), 175-211.

develop an electromotive force of that magnitude while for antenna currents of from 2,000–4,000 amperes at the usual frequencies it need not be larger than 30 feet in diameter.

In a more restricted meaning this wireless transmitter is one in which the Hertz-wave radiation is an entirely negligible quantity as compared with the whole energy, under which condition the damping factor is extremely small and an enormous charge is stored in the elevated capacity. Such a circuit may then be excited with impulses of any kind, even of low frequency and it will yield sinusoidal[12] and continuous oscillations like those of an alternator.

Taken in the narrowest significance of the term, however, it is a resonant transformer which, besides possessing these qualities, is accurately proportioned to fit the globe and its electrical constants and properties, by virtue of which design it becomes highly efficient and effective in the wireless transmission of energy. Distance is then absolutely eliminated, there being *no diminution in the intensity of the transmitted impulses*. It is even possible to make the actions *increase with the distance from the plant* according to an exact mathematical law.

This invention was one of a number comprised in my "World-System" of wireless transmission which I undertook to commercialize on my return to New York in 1900. As to the immediate purposes of my enterprise, they were clearly outlined in a technical statement of that period from which I quote:

> The 'World-System' has resulted from a combination of several original discoveries made by the inventor in the course of long continued research and experimentation. It makes possible not only the instantaneous and precise wireless transmission of any kind of signals, messages, or characters, to all parts of the world, but also the inter-connection of

---

[12] Having the form of a sine curve (a curve representing periodic oscillations of constant amplitude as given by a sine function).

the existing telegraph, telephone, and other signal stations without any change in their present equipment. By its means, for instance, a telephone subscriber here may call up and talk to any other subscriber on the Globe. An inexpensive receiver, not bigger than a watch, will enable him to listen anywhere, on land or sea, to a speech delivered or music played in some other place, however distant. These examples are cited merely to give an idea of the possibilities of this great scientific advance, which annihilates distance and makes that perfect natural conductor, the Earth, available for all the innumerable purposes which human ingenuity has found for a line-wire. One far-reaching result of this is that any device capable of being operated through one or more wires (at a distance obviously restricted) can likewise be actuated, without artificial conductors and with the same facility and accuracy, at distances to which there are no limits other than those imposed by the physical dimensions of the Globe. Thus, not only will entirely new fields for commercial exploitation be opened up by this ideal method of transmission but the old ones vastly extended.

The 'World-System' is based on the application of the following important inventions and discoveries:

1. The **Tesla Transformer** — This apparatus is in the production of electrical vibrations as revolutionary as gunpowder was in warfare. Currents many times stronger than any ever generated in the usual ways, and sparks over one hundred feet long, have been produced by the inventor with an instrument of this kind.

2. The **Magnifying Transmitter** — This is Tesla's best invention, a peculiar transformer specially adapted to excite the Earth, which is in the transmission of electrical energy what the telescope is in astronomical observation. By the

use of this marvelous device he has already set up electrical movements of greater intensity than those of lightning and passed a current, sufficient to light more than two hundred incandescent lamps, around the Globe.

3. The **Tesla Wireless System** — This system comprises a number of improvements and is the only means known for transmitting economically electrical energy to a distance without wires. Careful tests and measurements in connection with an experimental station of great activity, erected by the inventor in Colorado, have demonstrated that power in any desired amount can be conveyed, clear across the Globe if necessary, with a loss not exceeding a few percent.

4. The **Art of Individualization** — This invention of Tesla's is to primitive 'tuning' what refined language is to unarticulated expression. It makes possible the transmission of signals or messages absolutely secret and exclusive both in the active and passive aspect, that is, non-interfering as well as non-interferable. Each signal is like an individual of unmistakable identity and there is virtually no limit to the number of stations or instruments which can be simultaneously operated without the slightest mutual disturbance.

5. The **Terrestrial Stationary Waves** — This wonderful discovery, popularly explained, means that the Earth is responsive to electrical vibrations of definite pitch just as a tuning fork to certain waves of sound. These particular electrical vibrations, capable of powerfully exciting the Globe, lend themselves to innumerable uses of great importance commercially and in many other respects.

The first 'World-System' power plant can be put in operation in nine months. With this power plant it will be practicable to attain electrical activities up to ten million horsepower and it is designed to serve for as many technical achievements as are possible without due expense. Among these the following may be mentioned:

1. The inter-connection of the existing telegraph exchanges or offices all over the world;
2. The establishment of a secret and non-interferable government telegraph service;
3. The inter-connection of all the present telephone exchanges or offices on the Globe;
4. The universal distribution of general news, by telegraph or telephone, in connection with the Press;
5. The establishment of such a 'World-System' of intelligence transmission for exclusive private use;
6. The inter-connection and operation of all stock tickers of the world;
7. The establishment of a 'World-System' of musical distribution, etc.;
8. The universal registration of time by cheap clocks indicating the hour with astronomical precision and requiring no attention whatever;
9. The world transmission of typed or handwritten characters, letters, checks, etc.;
10. The establishment of a universal marine service enabling the navigators of all ships to steer perfectly without compass, to determine the exact location, hour and speed, to prevent collisions and disasters, etc.;
11. The inauguration of a system of world-printing on land and sea;
12. The world reproduction of photographic pictures and all kinds of drawings or records.

MY INVENTIONS & other essays

I also proposed to make demonstrations in the wireless transmission of power on a small scale but sufficient to carry conviction. Besides these I referred to other and incomparably more important applications of my discoveries which will be disclosed at some future date.

A plant was built on Long Island with a tower 187 feet high, having a spherical terminal about 68 feet in diameter. These dimensions were adequate for the transmission of virtually any amount of energy. Originally only from 200 to 300 kilowatts were provided but I intended to employ later several thousand horsepower. The transmitter was to emit a wave complex of special characteristics and I had devised a unique method of telephonic control of any amount of energy.[13]

The tower was destroyed two years ago but my projects are being developed and another one, improved in some features, will be constructed. On this occasion I would contradict the widely circulated report that the structure was demolished by the Government which owing to war conditions, might have created prejudice in the minds of those who may not know that the papers, which thirty years ago conferred upon me the honor of American citizenship, are always kept in a safe, while my orders, diplomas, degrees, gold medals, and other distinctions are packed away in old trunks. If this report had a foundation I would have been refunded a large sum of money which I expended in the construction of the tower. On the contrary it was in the interest of the Government to preserve it, particularly as it would have made possible — to mention just one valuable result — the location of a submarine in any part of the world. My plant, services, and all my improvements have always been at the disposal of the officials and ever since the outbreak of the European conflict I have been working at a sacrifice on several

---

[13] Wardenclyffe Tower, also known as the Tesla Tower, located in the village of Shoreham, New York, was an early experimental wireless transmission station designed and built by Tesla. Today, the location has been purchased by a nonprofit group and will soon become a Tesla museum and science center.

The Magnifying Transmitter

inventions of mine relating to aerial navigation, ship propulsion, and wireless transmission which are of the greatest importance to the country. Those who are well informed know that my ideas have revolutionized the industries of the United States and I am not aware that there lives an inventor who has been, in this respect, as fortunate as myself especially as regards the use of his improvements in the war. I have refrained from publicly expressing myself on this subject before as it seemed improper to dwell on personal matters while all the world was in dire trouble.

I would add further, in view of various rumors which have reached me, that Mr. J. Pierpont Morgan did not interest himself with me in a business way but in the same large spirit in which he has assisted many other pioneers. He carried out his generous promise to the letter and it would have been most unreasonable to expect from him anything more. He had the highest regard for my attainments and gave me every evidence of his complete faith in my ability to ultimately achieve what I had set out to do. I am unwilling to accord to some small-minded and jealous individuals the satisfaction of having thwarted my efforts. These men are to me nothing more than microbes of a nasty disease. My project was retarded by laws of nature. The world was not prepared for it. It was too far ahead of time. But the same laws will prevail in the end and make it a triumphal success.[14]

---

[14] Here, we believe Tesla is either trying to save face or being overly optimistic (both?) as his decision to scale up the facility at Wardenclyffe and implement his ideas of wireless power transmission, which required additional funding, to better compete with Guglielmo Marconi's radio-based telegraph system was met with refusal by the project's primary backer, J.P. Morgan. Additional investment could not be found and the project was abandoned in 1906, never to become operational.

This photograph shows the famous Tesla Tower erected at Shoreham, Long Island, New York. The tower was dismantled at the outbreak of the First World War. It was 187 feet high. The spherical top was 68 feet in diameter. Note the huge size of the structure by comparing the two-story power plant in the rear.

This model shows how the Tesla Tower built on Long Island would have looked completed.

SPECIAL NOTICE

Last month we announced another special feature article by Mr. Tesla, which although made in good faith by us was not authorized by him. Due to very important duties of Mr. Tesla, it was impossible for him to furnish his historical article this month. An important historical article will appear in the August issue.

—Editor.
Electrical Experimenter
July 1919

To the Editor of the *Electrical Experimenter*:

I regret that owing to circumstances beyond my control it was deemed advisable to postpone the publication of the article which I had intended for your August number. But the objections to its appearance will be shortly removed and I shall be pleased to forward it in due course for embodiment in your October issue.

Irrespective of this, I think it well on this occasion to notify your readers, as a precaution, that I am not one of those who display the sign "Do it now" on their desks and office doors. My motto is "Do not do it now. Think it over."

Very truly yours,

N. Tesla

In this article, Dr. Tesla dwells on the future possibilities of his magnifying transmitter, especially in connection with the art of Telautomatics, which was first conceived by him and doubtless constitutes one of his most brilliant gifts to the world.

Tesla was first to build and successfully operate Automata in the form of boats steered and otherwise controlled by tuned wireless circuits and agents ensuring reliable action despite of all attempts to interfere.

But this was only the first step in the evolution of his invention. What he wanted was to produce machines capable of acting as though possessed of intelligence. It will be readily perceived that if Dr. Tesla has practically realized his conception, the world will witness a revolution in every field of endeavor. In particular will his inventions affect the art of warfare and the peace of the world.

Dr. Tesla dwells eloquently on a number of topics agitating the public mind, and this article of his is perhaps the most brilliant and absorbing he has written.

—Editor.
Electrical Experimenter
October 1919

VI
The Art of Telautomatics

No subject to which I have ever devoted myself has called for such concentration of mind and strained to so dangerous a degree the finest fibers of my brain as the system of which the Magnifying Transmitter is the foundation. I put all the intensity and vigor of youth in the development of the rotating field discoveries, but those early labors were of a different character. Although strenuous in the extreme, they did not involve that keen and exhausting discernment which had to be exercised in attacking the many puzzling problems of the wireless. Despite my rare physical endurance at that period the abused nerves finally rebelled and I suffered a complete collapse, just as the consummation of the long and difficult task was almost in sight.

Without doubt I would have paid a greater penalty later, and very likely my career would have been prematurely terminated, had not providence equipped me with a safety device, which has seemed to improve with advancing years and unfailingly comes into play when my forces are at an end. So long as it operates I am safe from danger, due to overwork, which threatens other inventors and, incidentally, I need no vacations which are indispensable to most people. When I am all but used up I simply do as the darkies, who "naturally fall asleep while white folks worry."[1] To venture a theory out of my sphere, the body probably accumulates little by little a definite quantity of some toxic agent and I sink into a nearly lethargic state which lasts half an hour to the minute. Upon awakening I have the sensation as though the events immediately preceding had occurred very long ago, and if I attempt to continue the interrupted train of thought I feel a veritable mental nausea. Involuntarily I then turn to other work and am surprised at the freshness of the mind and ease with which I overcome obstacles that had baffled me before. After weeks or months my passion for the temporarily abandoned invention returns and I invariably find answers to all the vexing questions with scarcely any effort.

In this connection I will tell of an extraordinary experience which may be of interest to students of psychology. I had produced a striking phenomenon with my grounded transmitter and was endeavoring to ascertain its true significance in relation to the currents propagated through the earth. It seemed a hopeless undertaking, and for more than a year I worked unremittingly, but in vain. This profound study so entirely absorbed me that I became forgetful of everything else, even of my undermined health. At last, as I was at the point of breaking down, nature applied the preservative inducing lethal sleep. Regaining my senses I realized with consternation that I was unable to visualize scenes from my life except those of infancy, the very first ones that had entered my consciousness. Curiously enough, these appeared before my vision with startling distinctness and afforded me welcome relief. Night after night, when retiring, I would think of them and more and more of my previous existence was revealed. The image of my mother was always the principal figure in the spectacle that slowly

---

[1] We Heathens were not expecting that! Tesla was likely joking; however, "darkies" is an offensive term denoting persons with black or dark skin.

unfolded, and a consuming desire to see her again gradually took possession of me. This feeling grew so strong that I resolved to drop all work and satisfy my longing. But I found it too hard to break away from the laboratory, and several months elapsed during which I had succeeded in reviving all the impressions of my past life up to the spring of 1892. In the next picture that came out of the mist of oblivion, I saw myself at the *Hotel de la Paix*[2] in Paris just coming to from one of my peculiar sleeping spells, which had been caused by prolonged exertion of the brain. Imagine the pain and distress I felt when it flashed upon my mind that a dispatch was handed to me at that very moment bearing the sad news that my mother was dying. I remembered how I made the long journey home without an hour of rest and how she passed away after weeks of agony! It was especially remarkable that during all this period of partially obliterated memory I was fully alive to everything touching on the subject of my research. I could recall the smallest details and the least significant observations in my experiments and even recite pages of text and complex mathematical formulae.

My belief is firm in a law of compensation. The true rewards are ever in proportion to the labor and sacrifices made. This is one of the reasons why I feel certain that of all my inventions, the Magnifying Transmitter will prove most important and valuable to future generations. I am prompted to this prediction not so much by thoughts of the commercial and industrial revolution which it will surely bring about, but of the humanitarian consequences of the many achievements it makes possible. Considerations of mere utility weigh little in the balance against the higher benefits of civilization. We are confronted with portentous problems which cannot be solved just by providing for our material existence, however abundantly. On the contrary, progress in this direction is fraught with hazards and

---

[2] InterContinental Paris Le Grand Hotel, originally Grand-Hôtel de la Paix, is a historic luxury hotel in Paris, France, which opened in 1862. The renowned Café de la Paix has been located on its ground floor since it opened.

perils not less menacing than those born from want and suffering. If we were to release the energy of atoms or discover some other way of developing cheap and unlimited power at any point of the globe this accomplishment, instead of being a blessing, might bring disaster to mankind in giving rise to dissension and anarchy which would ultimately result in the enthronement of the hated regime of force. The greatest goodwill comes from technical improvements tending to unification and harmony, and my wireless transmitter is preeminently such. By its means the human voice and likeness will be reproduced everywhere and factories driven thousands of miles from waterfalls furnishing the power; aerial machines will be propelled around the earth without a stop and the sun's energy controlled to create lakes and rivers for motive purposes and transformation of arid deserts into fertile land. Its introduction for telegraphic, telephonic and similar uses will automatically cut out the statics and all other interferences which at present impose narrow limits to the application of the wireless.

This is a timely topic on which a few words might not be amiss. During the past decade a number of people have arrogantly claimed that they had succeeded in doing away with this impediment. I have carefully examined all of the arrangements described and tested most of them long before they were publicly disclosed, but the finding was uniformly negative. A recent official statement from the U.S. Navy may, perhaps, have taught some beguilable news editors how to appraise these announcements at their real worth. As a rule the attempts are based on theories so fallacious that whenever they come to my notice I cannot help thinking in a lighter vein. Quite recently a new discovery was heralded, with a deafening flourish of trumpets, but it proved another case of a mountain bringing forth a mouse.

This reminds me of an exciting incident which took place years ago when I was conducting my experiments with currents of high frequency. Steve Brodie had just jumped off the Brooklyn Bridge.[3] The feat has been vulgarized since by imitators, but the first report electrified New York. I was very impressionable then and frequently spoke of the daring printer. On a hot afternoon I felt the necessity of refreshing myself and stepped into one of the popular thirty thousand institutions of this great city where a delicious twelve percent beverage[4] was served which can now be had only by making a trip to the poor and devastated countries of Europe. The attendance was large and not over-distinguished and a matter was discussed which gave me an admirable opening for the careless remark: "This is what I said when I jumped off the bridge." No sooner had I uttered these words than I felt like the companion of Timotheus in the poem of Schiller.[5] In an instant there was a pandemonium and a dozen voices cried: "It is Brodie!" I threw a quarter on the counter and bolted for the door but the crowd was at my heels with yells: "Stop, Steve!" which must have been misunderstood for many persons tried to hold me up as I ran frantically for my haven of refuge. By darting around corners I fortunately managed — through the medium of a fire-escape — to reach the laboratory where I threw off my coat, camouflaged myself as a hard-working blacksmith, and started the forge. But these precautions proved unnecessary; I had eluded my pursuers. For many years afterward, at night, when imagination turns into specters the trifling troubles of the day, I often thought, as I tossed on the bed, what my fate would have been had that mob caught me and found out that I was not Steve Brodie!

---

[3] Steve Brodie (1861–1901) was an American from Manhattan, New York City, who on July 23, 1886, claimed to have jumped off the Brooklyn Bridge and survived. The supposed jump, of which the veracity was disputed, gave Brodie publicity, a thriving saloon, and a career as a performer. Brodie's fame persisted long past his death, with Brodie portrayed in films and with the slang term "Brodie" — as in to "do a Brodie" — entering American vernacular, meaning to take a chance or a leap, specifically a suicidal one.
[4] On average, most wines are 12% ABV (alcohol by volume).
[5] *The Cranes of Ibycus* written in 1797 is a ballad by Johann Christoph Friedrich (von) Schiller (1759–1805), a German playwright, poet, and philosopher.

Now the engineer, who lately gave an account before a technical body of a novel remedy against statics based on a "heretofore unknown law of nature," seems to have been as reckless as myself when he contended that these disturbances propagate up and down, while those of a transmitter proceed along the earth. It would mean that a condenser, as this globe, with its gaseous envelope, could be charged and discharged in a manner quite contrary to the fundamental teachings propounded in every elemental textbook of physics. Such a supposition would have been condemned as erroneous, even in Franklin's time, for the facts bearing on this were then well known and the identity between atmospheric electricity and that developed by machines was fully established. Obviously, natural and artificial disturbances propagate through the earth and the air in exactly the same way, and both set up electromotive forces in the horizontal, as well as vertical, sense. Interference cannot be overcome by any such methods as were proposed. The truth is this: in the air the potential increases at the rate of about fifty volts per foot of elevation, owing to which there may be a difference of pressure amounting to twenty, or even forty thousand volts between the upper and lower ends of the antenna. The masses of the charged atmosphere are constantly in motion and give up electricity to the conductor, not continuously but rather disruptively, this producing a grinding noise in a sensitive telephonic receiver. The higher the terminal and the greater the space encompassed by the wires, the more pronounced is the effect, but it must be understood that it is purely local and has little to do with the real trouble.

In 1900, while perfecting my wireless system, one form of apparatus comprised four antennae. These were carefully calibrated to the same frequency and connected in multiple with the object of magnifying the action, in receiving from any direction. When I desired to ascertain the origin of the transmitted impulses, each diagonally situated pair was put in series with a primary coil energizing the detector circuit. In the former case the sound was loud in the telephone; in the latter it ceased, as expected, the two antennae neutralizing each other, but the true statics manifested themselves in both instances and I had to devise special preventives embodying different principles.

By employing receivers connected to two points of the ground, as suggested by me long ago, this trouble caused by the charged air, which is very serious in the structures as now built, is nullified and besides, the liability of all kinds of interference is reduced to about one-half, because of the directional character of the circuit. This was perfectly self-evident, but came as a revelation to some simple-minded wireless folks whose experience was confined to forms of apparatus that could have been improved with an ax, and they have been disposing of the bear's skin before killing him. If it were true that strays performed such antics, it would be easy to get rid of them by receiving without aerials. But, as a matter of fact, a wire buried in the ground which, conforming to this view, should be absolutely immune, is more susceptible to certain extraneous impulses than one placed vertically in the air. To state it fairly, a slight progress has been made, but not by virtue of any particular method or device. It was achieved simply by discarding the enormous structures, which are bad enough for transmission but wholly unsuitable for reception, and adopting a more appropriate type of receiver. As I pointed out in a previous article, to dispose of this difficulty for good, a radical change must be made in the system, and the sooner this is done the better.

It would be calamitous, indeed, if at this time when the art is in its infancy and the vast majority, not excepting even experts, have no conception of its ultimate possibilities, a measure would be rushed through the legislature making it a government monopoly.

This was proposed a few weeks ago by Secretary Daniels,[6] and no doubt that distinguished official has made his appeal to the Senate and House of Representatives with sincere conviction. But universal evidence unmistakably shows that the best results are always obtained in healthful commercial competition. There are, however, exceptional reasons why wireless should be given the fullest freedom of development. In the first place it offers prospects immeasurably greater and more vital to betterment of human life than any other invention or discovery in the history of man. Then again, it must be understood that this wonderful art has been, in its entirety, evolved here and can be called "American" with more right and propriety than the telephone, the incandescent lamp, or the aeroplane. Enterprising press agents and stock jobbers have been so successful in spreading misinformation that even so excellent a periodical as the *Scientific American*[7] accords the chief credit to a foreign country. The Germans, of course, gave us the Hertz-waves and the Russian, English, French, and Italian experts were quick in using them for signaling purposes. It was an obvious application of the new agent and accomplished with the old classical and unimproved induction coil — scarcely anything more than another kind of heliography. The radius of transmission was very limited, the results attained of little value, and the Hertz oscillations, as a means for conveying intelligence, could have been advantageously replaced by sound-waves, which I advocated in 1891. Moreover, all of these attempts were made three years after the basic principles of the wireless system, which is universally employed today, and its potent instrumentalities had been clearly described and developed in America. No trace of those Hertzian appliances and methods

---

[6] Josephus Daniels (1862–1948) was an American newspaper editor and publisher; for decades he controlled North Carolina's largest newspaper the Raleigh *News and Observer*. He was appointed by President Woodrow Wilson to serve as United States Secretary of the Navy from 1913 to 1921.

[7] In print since 1845, *Scientific American* is the oldest continuously published monthly magazine in the United States.

remains today. We have proceeded in the very opposite direction and what has been done is the product of the brains and efforts of citizens of this country. The fundamental patents have expired and the opportunities are open to all. The chief argument of the Secretary is based on interference. According to his statement, reported in the *New York Herald*[8] of July 29th, signals from a powerful station can be intercepted in every village of the world . In view of this fact, which was demonstrated in my experiments of 1900, it would be of little use to impose restrictions in the United States.

As throwing light on this point, I may mention that only recently an odd looking gentleman called on me with the object of enlisting my services in the construction of world transmitters in some distant land. "We have no money," he said, "but carloads of solid gold and we will give you a liberal amount." I told him that I wanted to see first what will be done with my inventions in America, and this ended the interview. But I am satisfied that some dark forces are at work, and as time goes on the maintenance of continuous communication will be rendered more difficult. The only remedy is a system immune against interruption. It has been perfected, it exists, and all that is necessary is to put it in operation.

The terrible conflict is still uppermost in the minds and perhaps the greatest importance will be attached to the Magnifying Transmitter as a machine for attack and defense, more particularly in connection with *Telautomatics*. This invention is a logical outcome of observations begun in my boyhood and continued throughout my life. When the first results were published the *Electrical Review* stated editorially that it would become one of the "most potent factors in the advance and civilization of mankind." The time is not distant when this prediction will be fulfilled. In 1898 and 1900 it was offered to the Government and might have been adopted were

---

[8] The *New York Herald* was a large-distribution newspaper based in New York City that existed between 1835 and 1924, when it was acquired by its smaller rival the *New-York Tribune* to form the *New York Herald Tribune*.

I one of those who would go to Alexander's shepherd when they want a favor from Alexander.[9] At that time I really thought that it would abolish war, because of its unlimited destructiveness and exclusion of the personal element of combat. But while I have not lost faith in its potentialities, my views have changed since.

War cannot be avoided until the physical cause for its recurrence is removed and this, in the last analysis, is the vast extent of the planet on which we live. Only through annihilation of distance in every respect, as the conveyance of intelligence, transport of passengers and supplies, and transmission of energy will conditions be brought about some day, insuring permanency of friendly relations. What we now want most is closer contact and better understanding between individuals and communities all over the earth, and the elimination of that fanatic devotion to exalted ideals of national egoism and pride which is always prone to plunge the world into primeval barbarism and strife. No league or parliamentary act of any kind will ever prevent such a calamity. These are only new devices for putting the weak at the mercy of the strong. I have expressed myself in this regard fourteen years ago, when a combination of a few leading governments — a sort of Holy Alliance — was advocated by the late Andrew Carnegie,[10] who may be fairly considered as the father of this idea, having given to it more publicity and impetus than anybody else prior to the efforts of the President. While it cannot be denied that such a pact might be of material advantage to some less fortunate peoples, it cannot attain the chief object sought. Peace can only come as a natural consequence of universal enlightenment and merging of races, and we are still far from this blissful realization.

---

[9] Based on our research, we could not identify an origin for this phrase, but we believe it is an allusion to Alexander the Great.

[10] Andrew Carnegie (1835–1919) was a Scottish-American industrialist and philanthropist. He led the expansion of the American steel industry in the late 19th century with his Carnegie Steel Company and became one of the richest Americans in history.

As I view the world of today, in the light of the gigantic struggle we have witnessed, I am filled with conviction that the interests of humanity would be best served if the United States remained true to its traditions and kept out of "entangling alliances." Situated as it is, geographically, remote from the theaters of impending conflicts, without incentive to territorial aggrandizement,[11] with inexhaustible resources and immense population thoroughly imbued with the spirit of liberty and right, this country is placed in a unique and privileged position. It is thus able to exert, independently, its colossal strength and moral force to the benefit of all, more judiciously and effectively, than as member of a league.

In one of these biographical sketches, published in the *Electrical Experimenter*, I have dwelt on the circumstances of my early life and told of an affliction which compelled me to unremitting exercise of imagination and self-observation. This mental activity, at first involuntary under the pressure of illness and suffering, gradually became second nature and led me finally to recognize that I was but an automaton devoid of free will in thought and action and merely responsive to the forces of the environment. Our bodies are of such complexity of structure, the motions we perform are so numerous and involved, and the external impressions on our sense organs to such a degree delicate and elusive that it is hard for the average person to grasp this fact. And yet nothing is more convincing to the trained investigator than the mechanistic theory of life which had been, in a measure, understood and propounded by Descartes three hundred years ago.[12] But in his time many important functions of our organism were unknown and, especially with respect

[11] To increase or make greater in power, influence, stature, or reputation.
[12] Often called the father of modern philosophy, René Descartes (1596–1650) was a French philosopher, mathematician, and scientist who invented analytical geometry, linking the previously separate fields of geometry and algebra. His best known philosophical statement is "cogito, ergo sum" ("I think, therefore I am"), found in his *Discourse on the Method* (1637) and *Principles of Philosophy* (1644).

to the nature of light and the construction and operation of the eye, philosophers were in the dark.

In recent years the progress of scientific research in these fields has been such as to leave no room for a doubt in regard to this view on which many works have been published. One of its ablest and most eloquent exponents is, perhaps, Felix Le Dantec,[13] formerly assistant of Pasteur.[14] Prof. Jacques Loeb[15] has performed remarkable experiments in heliotropism,[16] clearly establishing the controlling power of light in lower forms of organisms, and his latest book, *Forced Movements*, is revelatory.[17] But while men of science accept this theory simply as any other that is recognized, to me it is a truth which I hourly demonstrate by every act and thought of mine. The consciousness of the external impression prompting me to any kind of exertion, physical or mental, is ever present in my mind. Only on very rare occasions, when I was in a state of exceptional concentration, have I found difficulty in locating the original impulses.

The by far greater number of human beings are never aware of what is passing around and within them, and millions fall victims of disease and die prematurely just on this account. The commonest everyday occurrences appear to them mysterious and inexplicable. One may feel a sudden wave of sadness and rake his brain for an explanation when he might have noticed that it was caused

[13] Félix-Alexandre Le Dantec (1869–1917) was a French biologist and philosopher of science.
[14] Regarded as both the father bacteriology and microbiology, Louis Pasteur (1822–1895) was a French chemist and microbiologist renowned for his discoveries of the principles of vaccination, microbial fermentation, and pasteurization. His works are credited to saving millions of lives through the developments of vaccines for rabies and anthrax.
[15] Jacques Loeb (1859–1924) was a German-born American physiologist and biologist who became one of the most famous scientists in America. He was the model for the character of Max Gottlieb in Sinclair Lewis' 1925 Pulitzer-winning novel *Arrowsmith*.
[16] The directional growth of a plant in response to sunlight.
[17] *Forced Movements, Tropisms, and Animal Conduct* (1918).

The Art of Telautomatics 95

by a cloud cutting off the rays of the sun. He may see the image of a friend dear to him under conditions which he construes as very peculiar, when only shortly before he has passed him in the street or seen his photograph somewhere. When he loses a collar button he fusses and swears for an hour, being unable to visualize his previous actions and locate the object directly. Deficient observation is merely a form of ignorance and responsible for the many morbid notions and foolish ideas prevailing. There is not more than one out of every ten persons who does not believe in telepathy and other psychic manifestations, spiritualism, and communion with the dead, and who would refuse to listen to willing or unwilling deceivers.

Just to illustrate how deeply rooted this tendency has become even among the clearheaded American population, I may mention a comical incident. Shortly before the war, when the exhibition of my turbines in this city elicited widespread comment in the technical papers, I anticipated that there would be a scramble among manufacturers to get hold of the invention, and I had particular designs on that man from Detroit who has an uncanny faculty for accumulating millions. So confident was I that he would turn up someday, that I declared this as certain to my secretary and assistants. Sure enough, one fine morning a body of engineers from the Ford Motor Company presented themselves with the request of discussing with me an important project. "Didn't I tell you?" I remarked triumphantly to my employees, and one of them said, "You are amazing, Mr. Tesla; everything comes out exactly as you predict." As soon as these hard-headed men were seated I, of course, immediately began to extol the wonderful features of my turbine, when the spokesmen interrupted me and said, "We know all about this, but we are on a special errand. We have formed a psychological society for the investigation of psychic phenomena and we want you to join us in this undertaking." I suppose those engineers never knew how near they came to being fired out of my office.

ch. ends p. 104

Ever since I was told by some of the greatest men of the time, leaders in science whose names are immortal, that I am possessed of an unusual mind, I bent all my thinking faculties on the solution of great problems regardless of sacrifice. For many years I endeavored to solve the enigma of death, and watched eagerly for every kind of spiritual indication. But only once in the course of my existence have I had an experience which momentarily impressed me as supernatural. It was at the time of my mother's death. I had become completely exhausted by pain and long vigilance, and one night was carried to a building about two blocks from our home. As I lay helpless there, I thought that if my mother died while I was away from her bedside she would surely give me a sign. Two or three months before I was in London in company with my late friend, Sir William Crookes,[18] when spiritualism was discussed, and I was under the full sway of these thoughts. I might not have paid attention to other men, but was susceptible to his arguments as it was his epochal work on radiant matter, which I had read as a student, that made me embrace the electrical career. I reflected that the conditions for a look into the beyond were most favorable, for my mother was a woman of genius and particularly excelling in the powers of intuition. During the whole night every fiber in my brain was strained in expectancy, but nothing happened until early in the morning, when I fell in a sleep, or perhaps a swoon, and saw a cloud carrying angelic figures of marvelous beauty, one of whom gazed upon me lovingly and gradually assumed the features of my mother. The appearance slowly floated across the room and vanished, and I was awakened by an indescribably sweet song of many voices. In that instant a certitude, which no words can express, came upon me that my mother had just died. And that was true. I was unable to understand the tremendous weight of the painful

---

[18] Sir William Crookes (1832–1919) was a British chemist, physicist, and pioneer of vacuum tubes, inventing the Crookes tube in 1875. A foundational invention that eventually changed the whole of chemistry and physics.

knowledge I received in advance, and wrote a letter to Sir William Crookes while still under the domination of these impressions and in poor bodily health. When I recovered I sought for a long time the external cause of this strange manifestation and, to my great relief, I succeeded after many months of fruitless effort. I had seen the painting of a celebrated artist, representing allegorically one of the seasons in the form of a cloud with a group of angels which seemed to actually float in the air, and this had struck me forcefully. It was exactly the same that appeared in my dream, with the exception of my mother's likeness. The music came from the choir in the church nearby at the early mass of Easter morning, explaining everything satisfactorily in conformity with scientific facts.

This occurred long ago, and I have never had the faintest reason since to change my views on psychical and spiritual phenomena, for which there is absolutely no foundation. The belief in these is the natural outgrowth of intellectual development. Religious dogmas are no longer accepted in their orthodox meaning, but every individual clings to faith in a supreme power of some kind. We all must have an ideal to govern our conduct and insure contentment, but it is immaterial whether it be one of creed, art, science, or anything else, so long as it fulfills the function of a dematerializing force. It is essential to the peaceful existence of humanity as a whole that one common conception should prevail.

While I have failed to obtain any evidence in support of the contentions of psychologists and spiritualists, I have proved to my complete satisfaction the automatism of life, not only through continuous observations of individual actions, but even more conclusively through certain generalizations. These amount to a discovery which I consider of the greatest moment to human society, and on which I shall briefly dwell. I got the first inkling of this astounding truth when I was still a very young man, but for many years I interpreted what I noted simply as coincidences. Namely, whenever either myself or a person to whom I was attached, or a cause to which I was devoted, was hurt by others in a particular way, which might be best popularly characterized as the most unfair imaginable, I experienced a singular and undefinable pain which, for want of a better term, I have qualified as "cosmic," and shortly thereafter, and invariably, those who had inflicted it came to grief. After many such cases I confided this to a number of friends, who had the opportunity to convince themselves of the truth of the theory which I have gradually formulated and which may be stated in the following few words:

Our bodies are of similar construction and exposed to the same external influences. This results in likeness of response and concordance of the general activities on which all our social and other rules and laws are based. We are automata entirely controlled by the forces of the medium being tossed about like corks on the surface of the water, but mistaking the resultant of the impulses from the outside for free will. The movements and other actions we perform are always life preservative and though seemingly quite independent from one another, we are connected by invisible links. So long as the organism is in perfect order it responds accurately to the agents that prompt it, but the moment that there is some derangement in any individual, his self-preservative power is impaired. Everybody understands, of course, that if one becomes deaf, has his eyesight weakened, or his limbs injured, the chances for his continued existence are lessened. But this is also true, and perhaps more so, of certain defects in the brain which deprive the automaton, more or less, of that vital quality and cause it to rush into destruction. A very sensitive and observant being, with his highly developed mechanism all intact, and acting with precision in obedience to the changing conditions of the environment, is endowed with a transcending mechanical sense, enabling him to evade perils too subtle to be directly perceived. When he comes in contact with others whose controlling organs are radically faulty, that sense asserts itself and he feels the "cosmic" pain. The truth of this has been borne out in hundreds of instances and I am inviting other students of nature to devote attention to this subject, believing that through combined and systematic effort results of incalculable value to the world will be attained.

The idea of constructing an automaton, to bear out my theory, presented itself to me early but I did not begin active work until 1893, when I started my wireless investigations. During the succeeding two or three years a number of automatic mechanisms, to be actuated from a distance, were constructed by me and exhibited to visitors in my laboratory. In 1896, however, I designed a complete machine capable of a multitude of operations, but the consummation of my labors was delayed until late in 1897. This machine was illustrated and described in my article in *The Century Magazine* of June 1900,[19] and other periodicals of that time and, when first shown in the beginning of 1898, it created a sensation such as no other invention of mine has ever produced. In November, 1898, a basic patent on the novel art was granted to me, but only after the Examiner-in-Chief had come to New York and witnessed the performance, for what I claimed seemed unbelievable. I remember that when later I called on an official in Washington, with a view of offering the invention to the Government, he burst out in laughter upon my telling him what I had accomplished. Nobody thought then that there was the faintest prospect of perfecting such a device. It is unfortunate that in this patent, following the advice of my attorneys, I indicated the control as being effected through the medium of a single circuit and a well-known form of detector, for the reason that I had not yet secured protection on my methods and apparatus for individualization. As a matter of fact, my boats were controlled through the joint action of several circuits and interference of every kind

---

[19] Tesla, N. (1900, June). The Problem of Increasing Human Energy. *The Century Magazine*, 60(2), 175-211.

100   **MY INVENTIONS** & other essays

was excluded. Most generally I employed receiving circuits in the form of loops, including condensers, because the discharges of my high-tension transmitter ionized the air in the hall so that even a very small aerial would draw electricity from the surrounding atmosphere for hours. Just to give an idea, I found, for instance, that a bulb 12 inches in diameter, highly exhausted, and with one single terminal to which a short wire was attached, would deliver well on to one thousand successive flashes before all charge of the air in the laboratory was neutralized. The loop form of receiver was not sensitive to such a disturbance and it is curious to note that it is becoming popular at this late date. In reality it collects much less energy than the aerials or a long grounded wire, but it so happens that it does away with a number of defects inherent to the present wireless devices. In demonstrating my invention before audiences, the visitors were requested to ask any questions, however involved, and the automaton would answer them by signs. This was considered magic at that time but was extremely simple, for it was myself who gave the replies by means of the device.

One of the telautomatic boats (submersible) constructed by Tesla and exhibited by him in 1898. Controlled by wireless without aerials.

The Art of Telautomatics

At the same period another larger telautomatic boat was constructed, a photograph of which is shown at left. It was controlled by loops, having several turns placed in the hull, which was made entirely water-tight and capable of submergence. The apparatus was similar to that used in the first with the exception of certain special features I introduced as, for example, incandescent lamps which afforded a visible evidence of the proper functioning of the machine.

These automata, controlled within the range of vision of the operator, were, however, the first and rather crude steps in the evolution of the Art of Telautomatics as I had conceived it. The next logical improvement was its application to automatic mechanisms beyond the limits of vision and at great distance from the center of control, and I have ever since advocated their employment as instruments of warfare in preference to guns. The importance of this now seems to be recognized, if I am to judge from casual announcements through the press of achievements which are said to be extraordinary but contain no merit of novelty, whatever. In an imperfect manner it is practicable, with the existing wireless plants, to launch an aeroplane, have it follow a certain approximate course, and perform some operation at a distance of many hundreds of miles. A machine of this kind can also be mechanically controlled in several ways and I have no doubt that it may prove of some usefulness in war. But there are, to my best knowledge, no instrumentalities in existence today with which such an object could be accomplished in a precise manner. I have devoted years of study to this matter and have evolved means, making such and greater wonders easily realizable.

As stated on a previous occasion, when I was a student at college I conceived a flying machine quite unlike the present ones. The underlying principle was sound but could not be carried into practice for want of a prime-mover of sufficiently great activity. In recent

Tesla's new self-propelled aerial tel-automaton. Devoid of propeller, sustaining wings, and all other means of external control. Can attain a speed of 350 mph, and will reach a predetermined point a thousand miles away accurately within a few feet.

years I have successfully solved this problem and am now planning aerial machines devoid of sustaining planes, ailerons,[20] propellers and other external attachments, which will be capable of immense speeds and are very likely to furnish powerful arguments for peace in the near future. Such a machine, sustained and propelled entirely by reaction, is shown at left and is supposed to be controlled either mechanically or by wireless energy. By installing proper plants it will be practicable to project a missile of this kind into the air and drop it almost on the very spot designated, which may be thousands of miles away. But we are not going to stop at this. Telautomata will be ultimately produced, capable of acting as if possessed of their own intelligence, and their advent will create a revolution. As early as 1898 I proposed to representatives of a large manufacturing concern the construction and public exhibition of an automobile carriage which, left to itself, would perform a great variety of operations involving something akin to judgment. But my proposal was deemed chimerical[21] at that time and nothing came from it.

At present many of the ablest minds are trying to devise expedients for preventing a repetition of the awful conflict which is only theoretically ended and the duration and main issues of which I have correctly predicted in an article printed in *The Sun*[22] of December 20, 1914.[23] The proposed League is not a remedy but on the contrary, in the opinion of a number of competent men, may bring about results just the opposite. It is particularly regrettable that a punitive policy was adopted in framing the terms of peace, because a few

---
[20] A hinged surface in the trailing edge of an airplane wing, used to control lateral balance.
[21] In this context, illusory or impossible to achieve.
[22] *The Sun* (1833–1950) was considered a serious paper, like New York's two more successful broadsheets, *The New York Times* and the *New York Herald Tribune*.
[23] Based on our research, we were able to discover two mentions advertising this December 20th article in *The Sun* archives – one on December 17th (A11), and the other on December 18th (A9) – yet no article by Tesla was printed in the December 20th, 1914, edition of the paper.

years hence it will be possible for nations to fight without armies, ships or guns, by weapons far more terrible, to the destructive action and range of which there is virtually no limit. A city, at any distance whatsoever from the enemy, can be destroyed by him and no power on earth can stop him from doing so. If we want to avert an impending calamity and a state of things which may transform this globe into an inferno, we should push the development of flying machines and wireless transmission of energy without an instant's delay and with all the power and resources of the nation.

& other essays

The Effect of Statics on Wireless Transmission

A few statements regarding these phenomena, in response to a request of the *Electrical Experimenter*, may be useful at the present time in view of the increasing interest and importance of the subject.

The commercial application of the art has led to the construction of larger transmitters and multiplication of their number, greater distances had to be covered and it became imperative to employ receiving devices of ever increasing sensitiveness. All these and other changes have cooperated in emphasizing the trouble and seriously impairing the reliability and value of the plants. To such a degree has this been the case that conservative business men and financiers have come to look upon this method of conveying intelligence as one offering but very limited possibilities, and the Government has deemed it advisable to assume control. This unfortunate state of affairs, fatal to enlistment of capital and healthful competitive development, could have been avoided had electricians not remained to this day under the spell of a delusive theory and had the practical exploiters of this advance not permitted enterprise to outrun technical competence.

With the publication of Dr. Heinrich Hertz's classical researches it was an obvious inference that the dark rays investigated by him could be used for signaling purposes, as those of light in heliography, and the first steps in this direction were made with his apparatus which, in 1896, was found capable of actuating receivers at a distance of a few miles. Three years prior to this, however, in lectures before the Franklin Institute and National Electric Light Association, I had described a wireless system radically opposite to the Hertzian in principle inasmuch as it depended on currents conducted through the earth instead of on radiations propagated through the atmosphere, presumably in straight lines.

The apparatus then outlined by me consisted of a transmitter comprising a primary circuit excited from an alternator or equivalent source of electrical energy and a high potential secondary resonant circuit, connected with its terminals to ground and to an elevated capacity, and a similar tuned receiving circuit including the operative device. On that occasion I expressed myself confidently on the feasibility of flashing in this manner not only signals to any terrestrial distance but transmitting power in unlimited amounts for all sorts of industrial purposes. The discoveries made and experimental results attained I made with a wireless power-plant erected in 1899, some of which were disclosed in the *Century Magazine* of June, 1900,[1] and several U. S. patents subsequently granted to me have, I believe, borne-out strikingly my foresight. In the meantime the Hertzian arrangements were gradually modified, one feature after another being abandoned, so that now not a vestige of them can be found and my system of four tuned circuits has been universally adopted, not only in its fundamentals but in every detail as the "quenched sparks," "ticker," "tone wheel," high frequency and rotating field

---
[1] Tesla, N. (1900, June). The Problem of Increasing Human Energy. *The Century Magazine*, 60(2), 175-211.

The Effect of Statics...

alternators, forms of discharges and mercury breaks, frequency changers, coils, condensers, regulating methods and devices, etc. This fact would give me supreme satisfaction were it not that the engineers, misinterpreting the nature of the effects, are making installments so defective in construction and mode of operation as to preclude the possibility of the great realization which might be brought within easy reach by proper application of the underlying principles and one of which — the most desirable at present — is the complete elimination of all static and other interference.

During the past few years several emphatic announcements have been made that a perfect solution of this problem had been discovered, but it was manifest from a casual perusal of these publications that the experts were ignoring certain truths of vital bearing on the question, and so long as this was the case no such claim could be substantiated. I achieved early success by keeping these steadily in mind and applying my efforts from the outset in the right and correct scientific direction.

I may contribute to the clearness of the subject in answering a question which I have been asked by the Editors of the *Electrical Experimenter* with reference to the report contained in the last issue, that signals had been received around the globe, an achievement the practicability of which I have fully demonstrated by experiment eighteen years ago.

The question is, how can Hertz waves be conveyed to such a distance in view of the curvature of the earth? A few words will be sufficient to show the absurdity of the prevailing opinion propounded in text books.

We are living on a conducting globe surrounded by a thin layer of insulating air, above which is a rarefied and conducting atmosphere. If the earth is represented by a sphere of 12½" radius, then the layer which may be considered insulating for high frequency currents of great tension is less than 1/64 of an inch thick. It is held that the Hertz waves, emanating from a transmitter, get to the distant receiver by successive reflections. The utter impossibility of this will be evident when it is shown by a simple calculation that the amount of energy received, even if it could be collected in its totality, is infinitesimal and would not actuate the most sensitive instrument known were it magnified many million times. The fact is these waves have no perceptible influence on a receiver if situated at a much smaller distance. It should be remembered, moreover, that since the first attempts the wave lengths have been increased until those advocated by me were adopted, in which this form of radiation has been reduced to one-billionth.

When a circuit, connected to ground and to an elevated capacity oscillates, two effects separate and distinct are produced: Hertz waves are radiated in a direction at right angles to the axis of symmetry of the conductor, *and simultaneously a current is passed through the earth*. The former propagates with the speed of light, the latter with a velocity proportionate to the cosecant[2] of an angle which from the origin to the opposite point of the globe varies from zero to 180°. Expressed in words, at the start the speed is infinite and diminishes, first rapidly and then slowly until a quadrant is traversed when the current proceeds with the speed of light. From that region on the velocity gradually increases, becoming infinite at the opposite point of the globe. In a patent granted to me in April, 1905, I have summed up this law of propagation in the statement that the projections of all half waves on the axis of symmetry of movement are equal, which means that the successive half waves, though of different length, cover exactly the same area. In the near future many wonderful results will be obtained by taking advantage of this fact.

There is a vast difference between these two forms of wave movement in their bearing on the transmission. The Hertz waves represent energy which is radiated and unrecoverable. The current

---
[2] The ratio of the hypotenuse (in a right-angled triangle) to the side opposite an acute angle; the reciprocal of sine.

The Effect of Statics... 111

energy, on the other hand, is preserved and can be recovered theoretically, at least, in its entirety. If the experts will free themselves from the illusions under which they are laboring, they will find that to overcome static disturbances all that is needed is a properly constructed transmitter and receiver without any additional devices or preventives. I have, however, devised several forms of apparatus eliminating statics even in the present defective wireless installations in which they are magnified many times. Such a form of instrument which I have used successfully is shown in the photograph below. These phenomena have been studied by me for a number of years and I have found that there are nine or ten different causes tending to intensify them, and in due course I shall give a full description of the various improvements I have made, in the *Electrical Experimenter*. For the present I would only point out that in order to perfectly eliminate the static interference, it is indispensable to redesign the whole wireless apparatus as now employed. The sooner this is understood the better it will be for the further evolution of the Art.

Tesla's Static Eliminator, patented and used by him circa 1900.

In this original and revolutionizing discussion, Nikola Tesla gives us something really new to think about. First — Does the moon rotate on its axis? Second — Is the Franklin pointed lightning rod correct in theory and operation? Third — Do wireless signals fly through space by means of so-called Hertzian waves in the ether, or are they propagated through the earth at prodigious velocity by means of earth-bound oscillations? World-famous conundrums these — questions which have been answered in many ways by some of the greatest scientists. Dr. Tesla explains these three predominant scientific fallacies in a masterly way, so that everyone can understand them.

—Editor.
Electrical Experimenter
February 1919

Famous Scientific Illusions

The human brain, with all its wonderful capabilities and power, is far from being a faultless apparatus. Most of its parts may be in perfect working order, but some are atrophied, undeveloped, or missing altogether. Great men of all classes and professions — scientists, inventors, and hard-headed financiers — have placed themselves on record with impossible theories, inoperative devices, and unrealizable schemes. It is doubtful that there could be found a single work of any one individual free of error. There is no such thing as an infallible brain. Invariably, some cells or fibers are wanting or unresponsive, with the result of impairing judgment, sense of proportion, or some other faculty. A man of genius eminently practical, whose name is a household word, has wasted the best years of his life in a visionary undertaking. A celebrated physicist was incapable of tracing the direction of an electric current according to a childishly simple rule. The writer, who was known to recite entire volumes by heart, has never been able to retain in memory and recapitulate in their proper order the words designating the colors of the rainbow, and can only ascertain them after long and laborious thought, strange as it may seem.

Our organs of reception, too, are deficient and deceptive.

MY INVENTIONS & other essays

As a semblance of life is produced by a rapid succession of inanimate pictures, so many of our perceptions are but trickery of the senses, devoid of reality. The greatest triumphs of man were those in which his mind had to free itself from the influence of delusive appearances. Such was the revelation of Buddha that self is an illusion caused by the persistence and continuity of mental images; the discovery of Copernicus that, contrary to all observation, this planet rotates around the sun; the recognition of Descartes that the human being is an automaton, governed by external influence and the idea that the earth is spherical, which led Columbus to the finding of this continent. And though the minds of individuals supplement one another and science and experience are continually eliminating fallacies and misconceptions, much of our present knowledge is still incomplete and unreliable. We have sophisms[1] in mathematics which cannot be disproved. Even in pure reasoning, free of the shortcomings of symbolic processes, we are often arrested by doubt which the strongest intelligences have been unable to dispel. Experimental science itself, most positive of all, is not unfailing.

In the following, I shall consider three exceptionally interesting errors in the interpretation and application of physical phenomena which have for years dominated the minds of experts and men of science.

I. The Illusion of the Axial Rotation of the Moon.

It is well known since the discovery of Galileo that the moon, in traveling through space, always turns the same face toward the earth. This is explained by stating that while passing once around its mother-planet the lunar globe performs just one revolution on its axis. The spinning motion of a heavenly body must necessarily undergo modifications in the course of time, being either retarded

---
[1] Fallacious arguments, especially ones used deliberately to deceive.

Famous Scientific Illusions 115

by resistances internal or external, or accelerated owing to shrinkage and other causes. An unalterable rotational velocity through all phases of planetary evolution is manifestly impossible. What wonder, then, that at this very instant of its long existence our satellite should revolve exactly so, and not faster or slower. But many astronomers have accepted as a physical fact that such rotation takes place. It does not, but only appears so; it is an illusion, a most surprising one, too.

I will endeavor to make this clear by reference to **Fig. 1**, in which *E* represents the earth and *M* the moon. The movement through

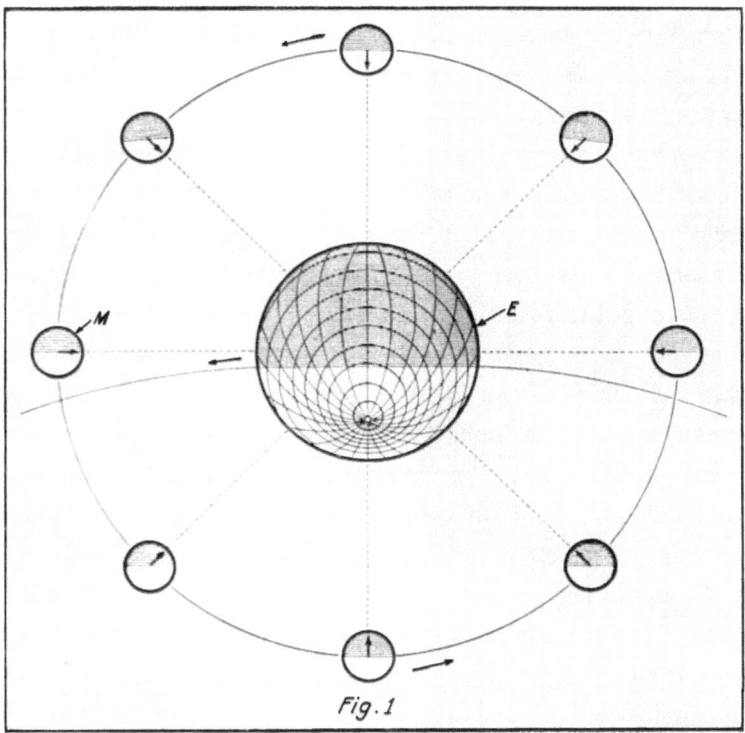

It is well know that the moon, *M*, always turns the same face toward the earth, *E*, as the black arrows indicate. The parallel rays from the sun illuminate the moon in its successive orbital positions as the unshaded semi-circles indicate. Bearing this in mind, do you believe the moon rotates on its own axis?

ch. ends p. 133

space is such that the arrow, firmly attached to the latter, always occupies the position indicated with reference to the earth. If one imagines himself as looking down on the orbital plane and follows the motion he will become convinced that the moon *does* turn on its axis as it travels around. But in this very act the observer will have deceived himself. To make the delusion complete let him take a washer similarly marked and supporting it rotatably in the center, carry it around a stationary object, constantly keeping the arrow pointing toward the latter. Though to his bodily vision the disk will revolve on its axis, such movement does not exist. He can dispel the illusion at once by holding the washer fixedly while going around. He will now readily see that the supposed axial rotation is only apparent, the impression being produced by successive changes of position in space.

But more convincing proofs can be given that the moon does not, and cannot revolve on its axis. With this object in view, attention is called to **Fig. 2**, in which both the satellite, M, and earth, E, are shown embedded in a solid mass, M_1, (indicated by stippling) and supposed to rotate so as to impact to the moon its normal translatory velocity. Evidently, if the lunar globe could rotate as commonly believed, this would be equally true of any other portion of mass M_1, as the sphere M_2, shown in dotted lines, and then the part common to both bodies would have to turn *simultaneously in opposite directions*. This can be experimentally illustrated in the manner suggested by using instead of one, two overlapping rotatable washers, as may be conveniently represented by circles M and M_2, and carrying them around a center as E, so that the plain and dotted arrows are always pointing toward the same center. No further argument is needed to demonstrate that the two gyrations cannot co-exist or even be pictured in the imagination and reconciled in a purely abstract sense.

The truth is, the so-called "axial rotation" of the moon is a phenomenon deceptive alike to the eye and mind and devoid of physical meaning. It has nothing in common with real mass revolution characterized by effects positive and unmistakable. Volumes have been written on the subject and many erroneous arguments advanced in support of the notion. Thus, it is reasoned, that if the planet did *not* turn on its axis it would expose the whole surface to terrestrial view; as only one-half is visible, it *must* revolve. The first statement is true but the logic of the second is defective, for it admits of only one alternative. The conclusion is not justified as the same appearance can also be produced in another way.

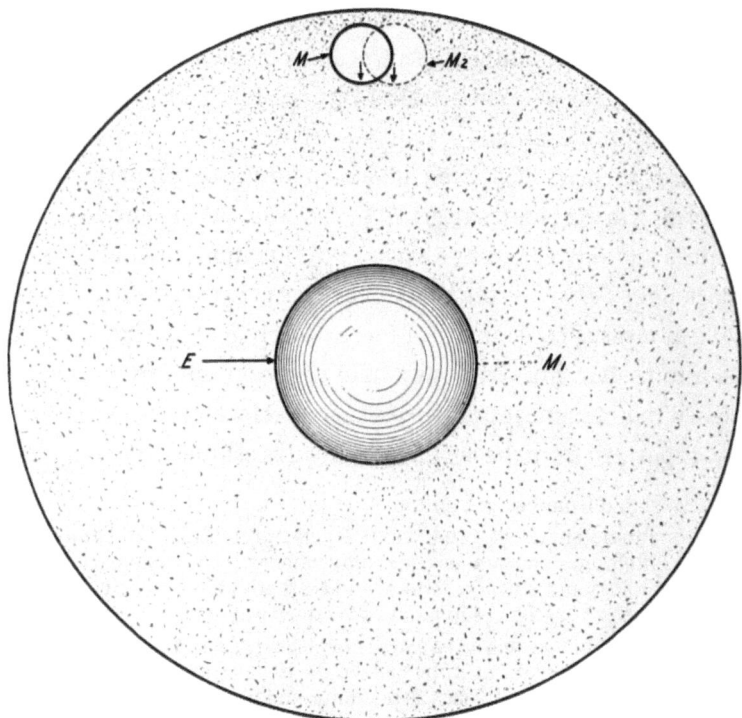

Fig. 2 — Tesla's conception of the rotation of the moon, *M*, around the earth, *E*; the moon, in this demonstration hypothesis, being considered embedded in a solid mass, M_1. If, as commonly believed, the moon rotates, this would be equally true for a portion of the mass M_2, and the part common to both bodies would turn simultaneously in "opposite" directions.

The moon does rotate, not on its own, **but about an axis passing through the center of the earth, the true and only one.**

The unfailing test of the spinning of a mass is, however, the existence of energy of motion. The moon is not possessed of such *vis viva*.[2] If it were the case then a revolving body as M_1 would contain mechanical energy other than that of which we have experimental evidence. Irrespective of this so exact a coincidence between the axial and orbital periods is, in itself, immensely improbable for this is not the permanent condition toward which the system is tending. Any axial rotation of a mass left to itself, retarded by forces external or internal, must cease. Even admitting its perfect control by tides the coincidence would still be miraculous. But when we remember that most of the satellites exhibit this peculiarity, the probability becomes infinitesimal.

Three theories have been advanced for the origin of the moon. According to the oldest suggested by the great German philosopher Kant,[3] and developed by Laplace[4] in his monumental treatise "Mécanique Céleste,"[5] the planets have been thrown off from larger central masses by centrifugal force. Nearly forty years ago Prof. George H. Darwin[6] in a masterful essay on tidal friction[7] furnished mathematical proofs, deemed unrefutable, that the moon had separated from the earth. Recently this established theory has

[2] Translated from Latin: a living.
[3] Immanuel Kant (1724–1804) was a German philosopher and one of the central Enlightenment thinkers; his comprehensive and systematic works in epistemology, metaphysics, ethics, and aesthetics have made him one of the most influential figures in modern Western philosophy.
[4] Pierre-Simon, marquis de Laplace (1749–1827) was a French scholar and polymath whose work was important to the development of engineering, mathematics, statistics, physics, astronomy, and philosophy.
[5] *Traité de mécanique céleste* ("Treatise of celestial mechanics") published in five volumes from 1798 to 1825.
[6] A son of Charles Darwin, Sir George Howard Darwin (1845–1912) was an English barrister (lawyer) and astronomer.
[7] *On the Bodily Tides of Viscous and Semi-Elastic Spheroids and on the Ocean Tides Upon a Yielding Nucleus* (1879).

been attacked by Prof. T. J. J. See[8] in a remarkable work on the "Evolution of the Stellar Systems,"[9] in which he propounds the view that centrifugal force was altogether inadequate to bring about the separation and that all planets, including the moon, have come from the depths of space and have been captured. Still a third hypothesis of unknown origin exists which has been examined and commented upon by Prof. W. H. Pickering[10] in *Popular Astronomy* of 1907,[11] and according to which the moon was torn from the earth when the later was partially solidified, this accounting for the continents which might not have been formed otherwise.

Undoubtedly planets and satellites have originated in both ways and, in my opinion, it is not difficult to ascertain the character of their birth. The following conclusions can be safely drawn:

1. A heavenly body thrown off from a larger one cannot rotate on its axis. The mass, rendered fluid by the combined action of heat and pressure, upon the reduction of the latter immediately stiffens, being at the same time deformed by gravitational pull. The shape becomes permanent upon cooling and solidification and the smaller mass continues to move about the larger one as though it were rigidly connected to it except for pendular swings or librations[12] due to varying

---

[8] Thomas Jefferson Jackson See (1866–1962) was an American astronomer whose promulgated theories in astronomy and physics were eventually disproven. His educational and professional career were dogged by conflict, including his attacks on relativity. He was fired from his position at two observatories, eventually serving out his professional years at a naval shipyard in California.

[9] *Researches on the Evolution of the Stellar Systems, Vol. 1* (1896) and *Vol. 2* (1910).

[10] William Henry Pickering (1858–1938) was an American astronomer who constructed and established several observatories or astronomical observation stations, notably including Percival Lowell's Observatory in Flagstaff, Arizona.

[11] Pickering, W.H. (1907, May). The Place of Origin of the Moon — The Volcanic Problem. *Popular Astronomy*, 15(5), 274–287.

[12] Apparent or real oscillations of the moon, by which parts near the edge of the disc that are often not visible from the earth sometimes come into view.

ch. ends p. 133

orbital velocity. Such motion precludes the possibility of axial rotation in the strictly physical sense. The moon has never spun around as is well demonstrated by the fact that the most precise measurements have failed to show any measurable flattening in form.

2. If a planetary body in its orbital movement turns the same side toward the central mass this is a positive proof that it has been separated from the latter and is a true satellite.

3. A planet revolving on its axis in its passage around another cannot have been thrown off from the same but must have been captured.

II. The Fallacy of Franklin's Pointed Lightning-Rod.

The display of atmospheric electricity has since ages been one of the most marvelous spectacles afforded to the sight of man. Its grandeur and power filled him with fear and superstition. For centuries he attributed lightning to agents god-like and supernatural and its purpose in the scheme of this universe remained unknown to him. Now we have learned that the waters of the ocean are raised by the sun and maintained in the atmosphere delicately suspended, that they are wafted to distant regions of the globe where electric forces assert themselves in upsetting the sensitive balance and causing precipitation, thus sustaining all organic life. There is every reason to hope that man will soon be able to control this life-giving flow of water and thereby solve many pressing problems of his existence.

Atmospheric electricity became of special scientific interest in Franklin's time. Faraday had not yet announced his epochal discoveries in magnetic induction but static frictional machines were already generally used in physical laboratories. Franklin's powerful mind at once leaped to the conclusion that frictional and atmospheric electricity were identical. To our present view this inference appears obvious, but in his time the mere thought of it was little short of blasphemy. He investigated the phenomena and argued that if they were of the same nature then the clouds could be drained of their charge exactly as the ball of a static machine, and in 1749 he indicated in a published memoir how this could be done by the use of pointed metal rods.

The earliest trials were made by Dalibard[13] in France, but Franklin himself was the first to obtain a spark by using a kite, in June, 1752. When these atmospheric discharges manifest themselves today in our wireless station we feel annoyed and wish that they would stop, but to the man who discovered them they brought tears of joy.

The lightning conductor in its classical form was invented by Benjamin Franklin in 1755 and immediately upon its adoption proved a success to a degree. As usual, however, its virtues were often exaggerated. So, for instance, it was seriously claimed that in the city of Pietermaritzburg (capital of Natal, South Africa) no lightning strikes occurred after the pointed rods were installed, although the storms were as frequent as before. Experience has shown that just the opposite is true. A modern city like New York, presenting innumerable sharp points and projections in good contact with the earth, is struck much more often than equivalent area of land. Statistical records, carefully compiled and published from time to time, demonstrate that the danger from lightning to property and life has been reduced to a small percentage by Franklin's invention, but the damage by fire amounts, nevertheless, to several million dollars annually. It is astonishing that this device, which has been in universal use for more than one century and a half, should be found to involve a gross fallacy in design and construction which impairs its usefulness and may even render its employment hazardous under certain conditions.

[13] Thomas-François Dalibard (1709–1778) was a French physicist and botanist. After translating Franklin's *Experiments and Observations on Electricity* into French, he successfully performed Franklin's proposed experiment using a 40-foot-tall metal rod at Marly-la-Ville in northern France on May 10, 1752.

122   **MY INVENTIONS** & other essays

For explanation of this curious fact I may first refer to **Fig. 3**, in which *s* is a metallic sphere of radius *r*, such as the capacity terminal of a static machine, provided with a sharply pointed pin of length *h*, as indicated. It is well known that the latter has the property of quickly dissipating the accumulated charge into the air. To examine this action in the light of present knowledge we may liken electric potential to temperature. Imagine that sphere *s* is heated to *T* degrees and that the pin or metal bar is a perfect conductor of heat so that its extreme end is at the same temperature *T*. Then if another sphere of larger radius, v_1, is drawn about the first and the temperature along this boundary is T_1, it is evident that there will be between the end of the bar and its surrounding a difference of temperature $T - T_1$, which will determine the outflow of heat. Obviously, if the adjacent medium was not affected by the hot sphere this temperature difference would be greater and more heat would be given off. Exactly so in the electric system.

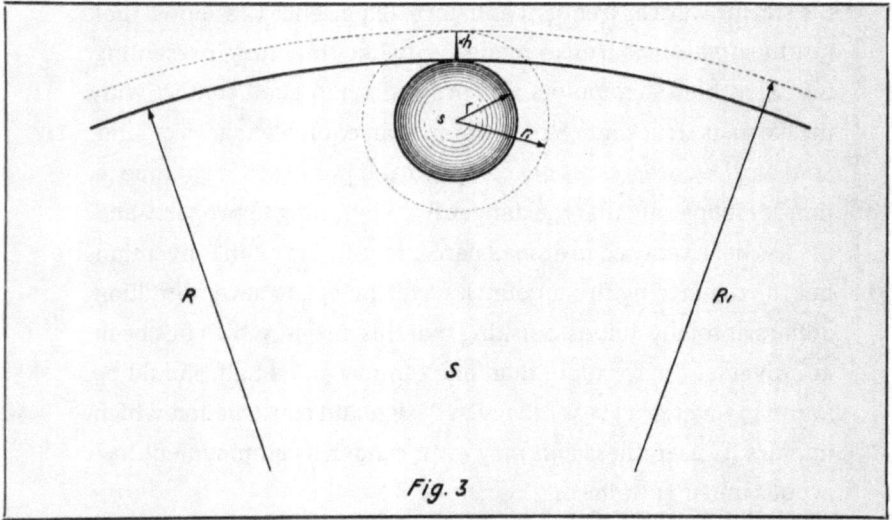

Diagram used to explain the fallacy of the Franklin Pointed Lightning Rod, and an analogy whereby Tesla shows in a clear manner how the charged sphere may for illustration be considered as heated to a high degree, and the heat allowed to escape at a known rate.

Let q be the quantity of the charge, then the sphere — and owing to its great conductivity also the pin — will be at the potential q/r. The medium around the point of the pin will be at the potential $q/r_1 = q/(r+h)$ and, consequently, the difference $q/r - q/(r+h) = qh/r(r+h)$. Suppose now that a sphere S of much larger radius $R = nr$ is employed containing a charge Q this difference of potential will be, analogously $Qh/R(R+h)$. According to elementary principles of electro-statics the potentials of the two spheres s and S will be equal if $Q = nq$ in which case $Qh/R(R+h) = nqh/nr(nr+h) = qh/r(nr+h)$. Thus the difference of potential between the point of the pin and the medium around the same will be smaller in the ratio $r+h/nr+h$ when the large sphere is used. In many scientific tests and experiments this important observation has been disregarded with the result of causing serious errors. Its significance is that the behavior of the pointed rod entirely depends on the linear dimensions of the electrified body. Its quality to give off the charge may be entirely lost if the latter is very large. For this reason, all points or projections on the surface of a conductor of such vast dimensions as the earth would be quite ineffective were it not for other influences. These will be elucidated with reference to **Fig. 4** on the following page, in which our artist of the Impressionist school has emphasized Franklin's notion that his rod was drawing electricity from the clouds. If the earth were not surrounded by an atmosphere which is generally oppositely charged it would behave, despite all its irregularities of surface, like a polished sphere. But owing to the electrified masses of air and cloud the distribution is greatly modified. Thus in **Fig. 4**, the positive charge of the cloud induces in the earth an equivalent *opposite* charge, the density at the surface of the latter diminishing with the cube of the distance from the static center of the cloud. A brush discharge is then formed at the point of the rod and the action Franklin anticipated takes place. In addition, the surrounding air is ionized and rendered conducting and, eventually, a bolt may hit the building or some other object in the vicinity. The virtue of the pointed end to dissipate the charge, which was uppermost in Franklin's mind is, however, infinitesimal. *Careful measurements show that it would take many years before the electricity stored in a single cloud of moderate size would be drawn off or neutralized through such a lightning conductor.* The grounded rod has the quality of rendering harmless most of the strokes it receives, though occasionally the charge is diverted with damaging results. But, what is very important to note, it invites danger and hazard on account of the fallacy involved in its design. The sharp point which was thought advantageous and indispensable to its operation, is really a defect detracting considerably from the practical value of the device. I have produced a much improved form of lightning protector characterized by the employment of a terminal of considerable area and large radius of curvature which

Fig. 4 — Tesla explains the fallacy of the Franklin Pointed Lightning Rod, here illustrated, and shows that usually such a rod could not draw off the electricity in a single cloud in many years. The density of the dots indicates the intensity of the charges.

makes impossible undue density of the charge and ionization of the air.[14] *These protectors act as quasi-repellents and so far have never been struck though exposed a long time.* Their safety is experimentally demonstrated to greatly exceed that invented by Franklin. By their use property worth millions of dollars, which is now annually lost, can be saved.

---

[14] Refer to pp. 205–209 wherein Tesla's form of non-pointed lightning rod is fully described and illustrated.

III. The Singular Misconception of the Wireless.

To the popular mind this sensational advance conveys the impression of a single invention but in reality it is an art, the successful practice of which involves the employment of a great many discoveries and improvements. I viewed it as such when I undertook to solve wireless problems and it is due to this fact that my insight into its underlying principles was clear from their very inception.

In the course of development of my induction motors it became desirable to operate them at high speeds and for this purpose I constructed alternators of relatively high frequencies. The striking behavior of the currents soon captivated my attention and in 1889 I started a systematic investigation of their properties and the possibilities of practical application. The first gratifying result of my efforts in this direction was the transmission of electrical energy through *one wire* without return, of which I gave demonstrations in my lectures and addresses before several scientific bodies here and abroad in 1891 and 1892. During that period, while working with my oscillation transformers and dynamos of frequencies up to 200,000 cycles per second, the idea gradually took hold of me that the earth might be used in place of the wire, thus dispensing with artificial conductors altogether. The immensity of the globe seemed an unsurmountable obstacle but after a prolonged study of the subject I became satisfied that the undertaking was rational, and in my lectures before the Franklin Institute and National Electric Light Association early in 1893 I gave the outline of the system I had conceived. In the latter part of that year, at the Chicago World's Fair, I had the good fortune of meeting Prof. Helmholtz to whom I explained my plan, illustrating it with experiments. On that occasion I asked the celebrated physicist for an expression of opinion on the feasibility of the scheme. He stated unhesitatingly that it was practicable, provided I could perfect apparatus capable of putting it into effect but this, he anticipated, would be extremely difficult to accomplish.

I resumed the work very much encouraged and from that date to 1896 advanced slowly but steadily, making a number of improvements the chief of which was my system of *concatenated tuned circuits*[15] and method of regulation, now universally adopted. In the summer of 1897 Lord Kelvin happened to pass through New York and honored me by a visit to my laboratory where I entertained him with demonstrations in support of my wireless theory. He was fairly carried away with what he saw but, nevertheless, condemned my project in emphatic terms, qualifying it as something impossible, "an illusion and a snare." I had expected his approval and was pained and surprised. But the next day he returned and gave me a better opportunity for explanation of the advances I had made and of the true principles underlying the system I had evolved. Suddenly he remarked with evident astonishment: "Then you are not making use of Hertz waves?"

"Certainly not," I replied, *"these are radiations.* No energy could be economically transmitted to a distance by any such agency. In my system the process is one of true conduction which, theoretically, can be effected at the greatest distance without appreciable loss." I can never forget the magic change that came over the illustrious philosopher the moment he freed himself from that erroneous impression. The skeptic who would not believe was suddenly transformed into the warmest of supporters. He parted from me not only thoroughly convinced of the scientific soundness of the idea but strongly expressed his confidence in its success. In my exposition to him I resorted to the following mechanical analogues of my own and the Hertz wave system.

Imagine the earth to be a bag of rubber filled with water, a small quantity of which is periodically forced in and out of the same by

---

[15] To concatenate is to link items together in a series.

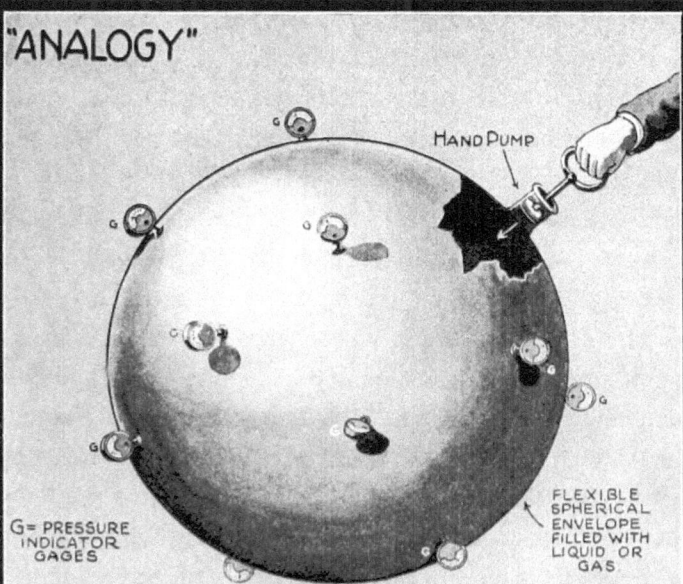
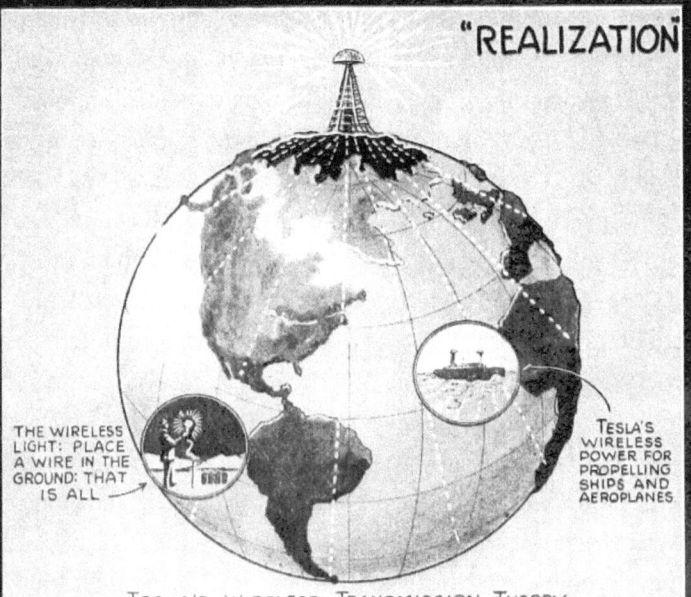

means of a reciprocating pump, as illustrated at right. If the strokes of the latter are effected in intervals of more than one hour and forty-eight minutes, sufficient for the transmission of the impulse through the whole mass, the entire bag will expand and contract and corresponding movements will be imparted to pressure gages or movable pistons with the same intensity, irrespective of distance. By working the pump faster, shorter waves will be produced which, on reaching the opposite end of the bag, may be reflected and give rise to stationary nodes and loops, but in any case, the fluid being incompressible, its enclosure perfectly elastic, and the frequency of oscillations not very high, the energy will be economically transmitted and very little power consumed so long as no work is done in the receivers. This is a crude but correct representation of my wireless system in which, however, I resort to various refinements. Thus, for instance, the pump is made part of a resonant system of great inertia, enormously magnifying the force of the impressed impulses. The receiving devices are similarly conditioned and in this manner the amount of energy collected in them vastly increased.

The Hertz wave system is in many respects the very opposite of this. To explain it by analogy, the piston of the pump is assumed to vibrate to and fro at a terrific rate and the orifice through which the fluid passes in and out of the cylinder is reduced to a small hole. There is scarcely any movement of the fluid and almost the whole work performed results in the production of radiant heat, of which an infinitesimal part is recovered in a remote locality. However incredible, it is true that the minds of some of the ablest experts have been from the beginning, and still are, obsessed by this monstrous idea, and so it comes that the true wireless art, to which I laid the foundation in 1893, has been retarded in its development for twenty years. This is the reason why the "statics" have proved unconquerable, why the wireless shares are of little value, and why the Government has been compelled to interfere.

We are living on a planet of well-nigh inconceivable dimensions, surrounded by a layer of insulating air above which is a rarefied and conducting atmosphere (**Fig. 5**). This is providential, for if all the air were conducting the transmission of electrical energy through the natural media would be impossible. My early experiments have shown that currents of high frequency and great tension readily pass through an atmosphere but moderately rarefied, so that the insulating stratum is reduced to a small thickness as will be evident by inspection of **Fig. 6**, in which a part of the earth and its gaseous envelope is shown to scale. If the radius of the sphere is 12½", then the non-conducting layer is only 1/64" thick and it will be obvious that the Hertzian rays cannot traverse so thin a crack between two conducting surfaces for any considerable distance, without being

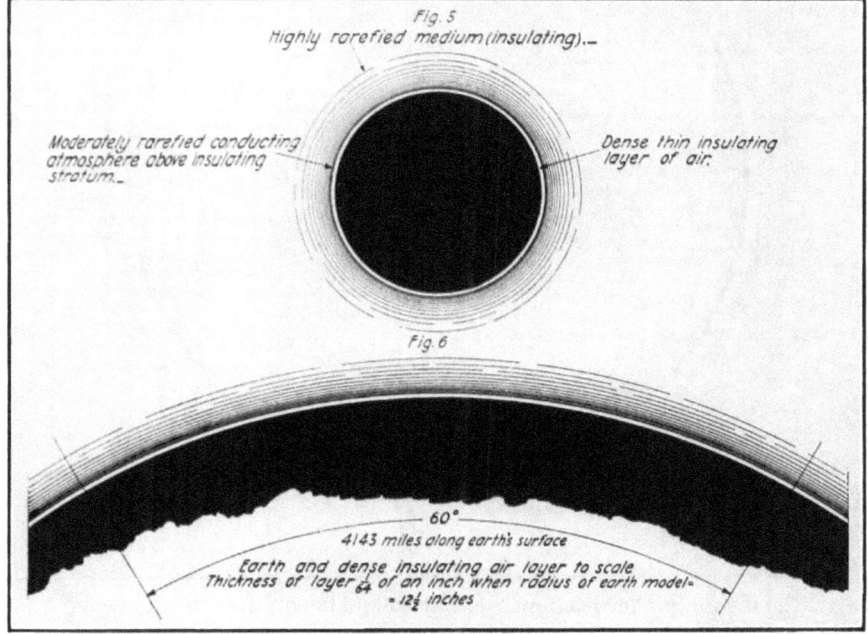

A section of the earth and its atmospheric envelope drawn to scale. It is obvious that the Hertzian rays cannot traverse so thin a crack between two conducting surfaces for any considerable distance, without being absorbed, says Dr. Tesla, in discussing the Ether Space Wave Theory.

absorbed. The theory has been seriously advanced that these radiations pass around the globe by *successive reflections*, but to show the absurdity of this suggestion reference is made to **Fig.** 7 in which this process is diagrammatically indicated. Assuming that there is no refraction, the rays, as shown on the right of the figure, would travel along the sides of a polygon drawn around the solid, and inscribed into the conducting gaseous boundary in which case the length of the side would be about 400 miles. As one-half the circumference of

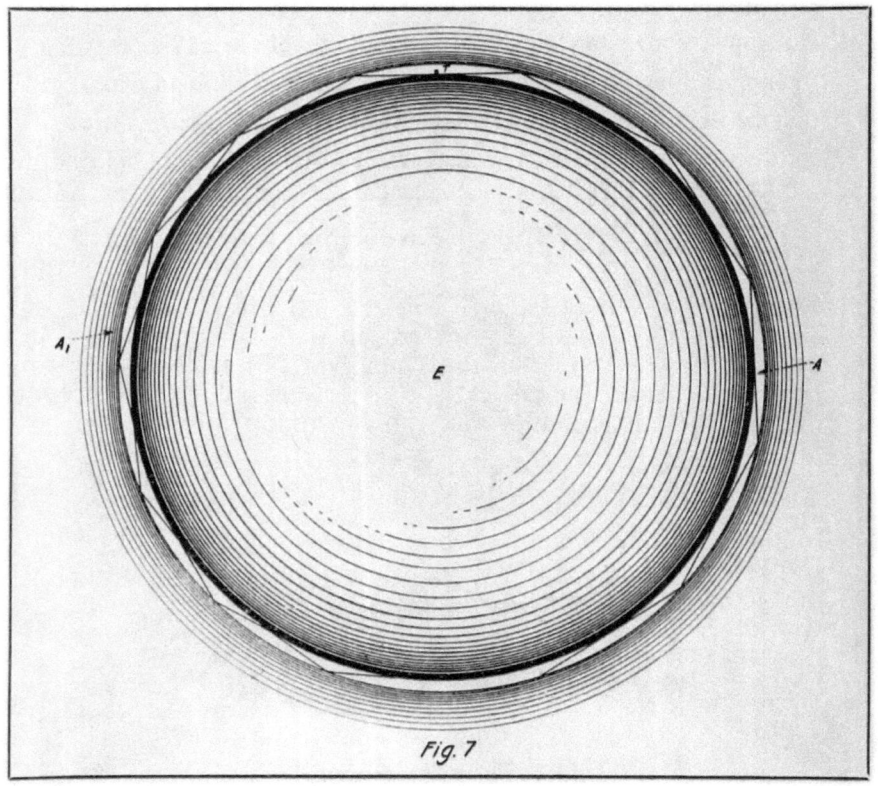

Fig. 7

The theory has been seriously advanced and taught that the radio ether wave oscillations pass around the earth by successive reflections, as here shown. The efficiency of such a reflector cannot be more than 25 percent; the amount of energy recoverable in a 12,000-mile transmission being one hundred and fifteen billionth part of one watt, with 1,000 kilowatts at the transmitter.

the earth is approximately 12,000 miles long there will be, roughly, thirty deviations. The efficiency of such a reflector cannot be more than 25 percent, so that if none of the energy of the transmitter were lost in other ways, the part recovered would be measured by the fraction $(1/4)^{30}$. Let the transmitter radiate Hertz waves at the rate of 1,000 kilowatts. Then about *one hundred and fifteen billionth part of one watt* is all that would be collected in a *perfect* receiver. In truth, the reflections would be much more numerous as shown on the left of the figure, and owing to this and other reasons, on which it is unnecessary to dwell, the amount recovered would be a vanishing quantity.

Consider now the process taking place in the transmission by the instrumentalities and methods of my invention. For this purpose attention is called to **Fig. 8** on the following page, which gives an idea of the mode of propagation of the current waves and is largely self-explanatory. The drawing represents a solar eclipse with the shadow of the moon just touching the surface of the earth at a point where the transmitter is located. As the shadow moves downward it will spread over the earth's surface, first with infinite and then gradually diminishing velocity until at a distance of about 6,000 miles it will attain its true speed in space. From there on it will proceed with increasing velocity, reaching infinite value at the opposite point of the globe. It hardly need be stated that this is merely an illustration and not an accurate representation in the astronomical sense.

The exact law will be readily understood by reference to **Fig. 9** on the following page, in which a transmitting circuit is shown connected to earth and to an antenna. The transmitter being in action, two effects are produced: Hertz waves pass through the air, and a current traverses the earth. The former propagate with the speed of light and their energy is *unrecoverable* in the circuit. The latter proceeds with the speed varying as the cosecant of the angle which a radius drawn from any point under consideration forms with the axis of symmetry of the waves. At the origin the speed is infinite but gradually diminishes until a quadrant is traversed, when the velocity is that of light. From there on it again increases, becoming infinite at the antipole. Theoretically the energy of this current is *recoverable* in its entirety, in properly attuned receivers.

Some experts, whom I have credited with better knowledge, have for years contended that my proposals to transmit power without wires are sheer nonsense but I note that they are growing more cautious every day. The latest objection to my system is found in the

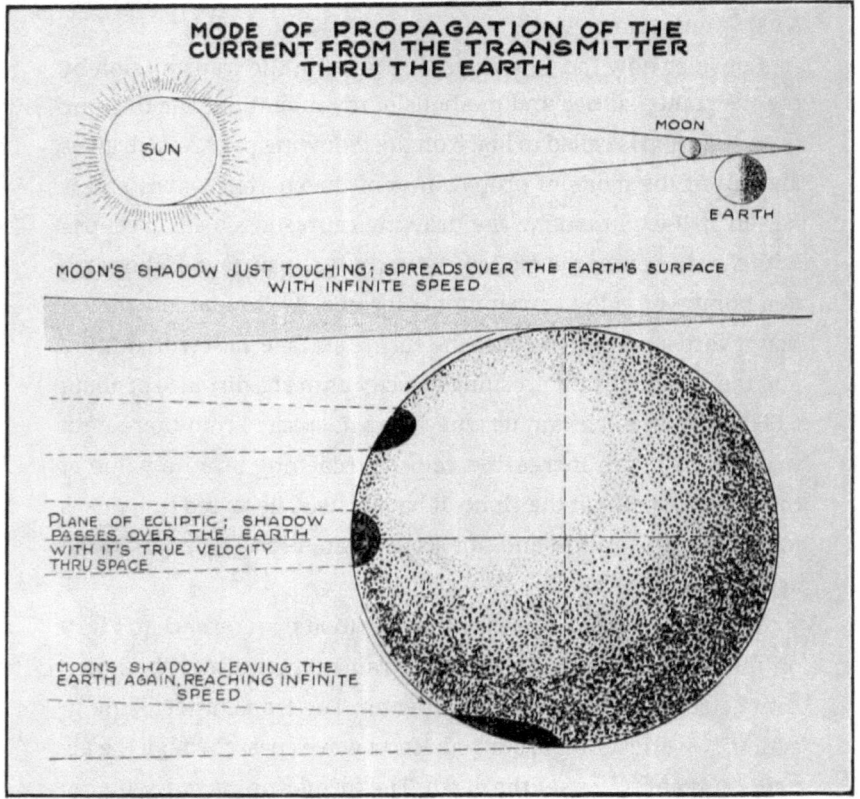

Fig. 8 — This diagram illustrates how, during a solar eclipse, the moon's shadow passes over the earth with changing velocity, and should be studied in connection with **Fig. 9**. The shadow moves downward with infinite velocity at first, then with its true velocity through space, and finally with infinite velocity again.

cheapness of gasoline. These men labor under the impression that the energy flows in all directions and that, therefore, only a minute amount can be recovered in any individual receiver. But this is far from being so. The power is conveyed in only one direction, from the transmitter to the receiver, and none of it is lost elsewhere. It is perfectly practicable to recover at any point of the globe energy enough for driving an airplane, or a pleasure boat, or for lighting a dwelling. I am especially sanguine in regard to the lighting of isolated places and believe that a more economical and convenient method can hardly be devised. The future will show whether my foresight is as accurate now as it has proved heretofore.

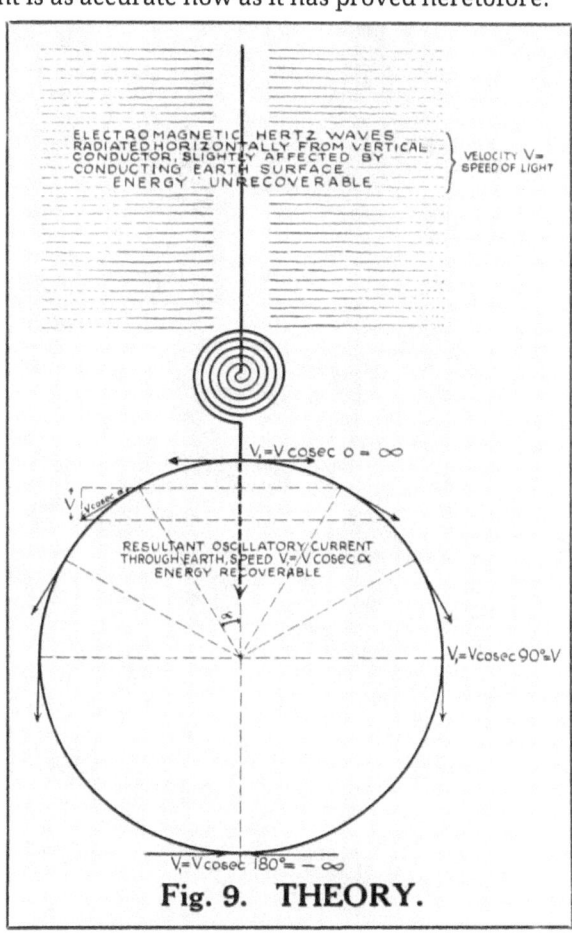

Fig. 9. THEORY.

EDITORIAL

It will come as a profound shock to all wireless enthusiasts, scientific and amateur alike, that their present-day notions on wireless are totally erroneous and not based upon actual facts. For years we clung to the theory that a wireless message radiates from the aerial wires of the sending station and speeds over the surface of the earth through the ether toward the receiving station. We thought that we were sending out pure Hertzian waves from our transmitters. We thought that we received these waves over the aerial wires of our receiving station. All of these theories are wrong and will be relegated shortly into the past along with the early notion that the earth stood still, while sun, moon, and stars revolved around it.

Remain only the physical facts that we did send and did receive messages without wires — but they are not sent by means of pure Hertz waves, nor do they go by way of the ether as radiations.

In a highly illuminating article printed elsewhere in this issue,[1] Nikola Tesla explodes all of our present orthodox views as to wireless propagation and makes it clear that the earth is the sole medium through which our wireless impulses travel, in the form of true

---
[1] A reference to Tesla's article "Famous Scientific Illusions," but we Heathens felt this editorial was better placed as preface to his next article "The True Wireless," in which he expands on his thoughts outlined in the previous.

conduction. Particularly does this hold true for long distance messages: Here we are sending out a compound impulse three quarters of which is a galvanic current, traveling through the conducting earth, the other quarter or less is in the form of Hertz waves, going by way of the ether. This explains why we can send signals to airplanes and vice versa; but even here we probably have to do not with pure Hertz waves; it is almost certain that we have capacity-inductive effects as well.

Tesla maintaining that there can be no long distance effects by radiations transmitted through the ether, but rather only by currents through the earth, it follows that in his opinion all our radio apparatus is designed and operated faultily. Indeed, this is not a brand new idea of the famous inventor. He has been preaching it ever since he took out his first patents and described his system in 1893 — long before Marconi thought of wireless. But he was preaching to a stone deaf scientific world.

But how simple it all becomes when we stop to apply a little reason and logic to Tesla's claims. For instance, we can send radio impulses three to five times as far over salt water as over land. Why? Simply because the impulses go *through* the water, which is a much better conductor than earth alone. If we were sending pure Hertzian waves, why do we connect one wire at both sending and receiving station to the ground? Hertz never dreamt of such a thing. If you are still unconvinced that the earth is the chief medium of transmission, disconnect your ground wires entirely and try to send and receive. Now you may work with Hertz waves, but the distances you can bridge will be pitifully small.

Already Tesla's logic is filtering into our radio scientists' minds. *All the big stations are beginning to scrap their towers and aerial wires, at least for receiving. They now bury their "aerial" wires in the ground,* and lo! They can receive signals twice as far as before. Incredible, but it is being done every day. And — wonders upon wonders — how we will laugh at our present and past blindness — *the static interference is practically gone* the minute we pull our aerial wires down and bury them! Static Electricity? There never was a reason for having the bugaboo, for there is no "static" in the ground.

But Tesla goes much farther. In time he will show the world wireless power transmission effected *not by ether waves but by current through the earth*, which is a first rate conductor. Like all big things, the problem is simple. At some point on the globe he will erect a station powerful enough to charge the whole earth with electricity — and keep it charged. To do this we need about 10,000 kilowatts. Then at *any* point on the globe the current can be tapped by means of suitable apparatus. Like a bell ringing transformer, *connected* to your supply line, no current is consumed unless you close the secondary circuit. Tesla's world wireless works just that way. *No current is consumed till it is tapped at the distant receiving station.*

Hugo Gernsback
Electrical Experimenter
February 1919

In this remarkable and complete story of his discovery of the "True Wireless" and the principles upon which transmission and reception, even in the present day systems, are based, Dr. Nikola Tesla shows us that he is indeed the "Father of the Wireless." To him the Hertz wave theory is a delusion; it looks sound from certain angles, but the facts tend to prove that it is hollow and empty. He convinces us that the real Hertz waves are blotted out after they have traveled but a short distance from the sender. It follows, therefore, that the measured antenna current is no indication of the effect, because only a small part of it is effective at a distance. The limited activity of pure Hertz wave transmission and reception is here clearly explained, besides showing definitely that in spite of themselves, the radio engineers of today are employing the original Tesla tuned oscillatory system. He shows by examples with different forms of aerials that the signals picked up by the instruments must actually be induced by earth currents — not etheric space waves. Tesla also disproves the "Heaviside layer" theory from his personal observations and tests.

—Editor.
Electrical Experimenter
May 1919

The True Wireless

Ever since the announcement of Maxwell's electro-magnetic theory,[1] scientific investigators all the world over had been bent on its experimental verification. They were convinced that it would be done and lived in an atmosphere of eager expectancy, unusually favorable to the reception of any evidence to this end. No wonder then that the publication of Dr. Heinrich Hertz's results caused a thrill as had scarcely ever been experienced before. At that time I was in the midst of pressing work in connection with the commercial introduction of my system of power transmission, but, nevertheless, caught the fire of enthusiasm and fairly burned with desire to behold the miracle with my own eyes. Accordingly, as soon as I had freed myself of these imperative duties and resumed research work in my laboratory on Grand Street, New York, I began, parallel with high frequency alternators, the construction of several forms of apparatus with the object of exploring the field opened up by Dr. Hertz. Recognizing the limitations of the devices he had employed, I concentrated my attention on the production of a powerful induction

---

[1] Named after the Scottish scientist James Clerk Maxwell (1831–1879), Maxwell's equations describe how electric and magnetic fields are generated by charges, currents, and changes of the fields.

coil but made no notable progress until a happy inspiration led me to the invention of the oscillation transformer. In the latter part of 1891, I was already so far advanced in the development of this new principle that I had at my disposal means vastly superior to those of the German physicist. All my previous efforts with Rhumkorf coils[2] had left me unconvinced, and in order to settle my doubts I went over the whole ground once more, very carefully, with these improved appliances. Similar phenomena were noted, greatly magnified in intensity, but they were susceptible of a different and more plausible explanation. I considered this so important that in 1892 I went to Bonn, Germany, to confer with Dr. Hertz in regard to my observations. He seemed disappointed to such a degree that I regretted my trip and parted from him sorrowfully. During the succeeding years I made numerous experiments with the same object, but the results were uniformly negative. In 1900, however, after I had evolved a wireless transmitter which enabled me to obtain electro-magnetic activities of many millions of horsepower, I made a last desperate attempt to prove that the disturbances emanating from the oscillator were ether vibrations akin to those of light, but met again with utter failure. For more than eighteen years I have been reading treatises, reports of scientific transactions, and articles on Hertz-wave telegraphy, to keep myself informed, but they have always impressed me like works of fiction.

The history of science shows that theories are perishable. With every new truth that is revealed we get a better understanding of Nature and our conceptions and views are modified. Dr. Hertz did not discover a new principle. He merely gave material support to a hypothesis which had been long ago formulated. It was a perfectly well-established fact that a circuit, traversed by a periodic current,

---

[2] Heinrich Daniel Rühmkorff (1803–1877) was a German instrument maker who commercialized the induction coil or spark coil (also referred to as the Ruhmkorff coil), which is a type of electrical transformer used to produce high-voltage pulses from a low-voltage direct current (DC) supply.

emitted some kind of space waves, but we were in ignorance as to their character. He apparently gave an experimental proof that they were transversal vibrations in the ether. Most people look upon this as his great accomplishment. To my mind it seems that his immortal merit was not so much in this as in the focusing of the investigators' attention on the processes taking place in the ambient medium. The Hertz-wave theory, by its fascinating hold on the imagination, has stifled creative effort in the wireless art and retarded it for twenty-five years. But, on the other hand, it is impossible to over-estimate the beneficial effects of the powerful stimulus it has given in many directions.

As regards signaling without wires, the application of these radiations for the purpose was quite obvious. When Dr. Hertz was asked whether such a system would be of practical value, he did not think so, and he was correct in his forecast. The best that might have been expected was a method of communication similar to the heliographic and subject to the same or even greater limitations.

In the spring of 1891, I gave my demonstrations with a high frequency machine before the American Institute of Electrical Engineers at Columbia College, which laid the foundation to a new and far more promising departure. Although the laws of electrical resonance were well known at that time and my lamented friend, Dr. John Hopkinson,[3] had even indicated their specific application to an alternator in the Proceedings of the Institute of Electrical Engineers, London, Nov. 13, 1889, nothing had been done toward the practical use of this knowledge and it is probable that those experiments of mine were the first public exhibition with resonant circuits, more particularly of high frequency. While the spontaneous success of

---

[3] John Hopkinson (1849–1898) was a British physicist, electrical engineer, Fellow of the Royal Society, and was twice the President of the Institution of Electrical Engineers (IEE, now the IET). He invented the three-wire (three-phase) system for the distribution of electrical power, for which he was granted a patent in 1882. Hopkinson's law, the magnetic counterpart to Ohm's law, is named after him.

my lecture was due to spectacular features, its chief import was in showing that all kinds of devices could be operated through a single wire without return. This was the initial step in the evolution of my wireless system. The idea presented itself to me that it might be possible, under observance of proper conditions of resonance, to transmit electric energy through the earth, thus dispensing with all artificial conductors. Anyone who might wish to examine impartially the merit of that early suggestion must not view it in the light of present day science. I only need to say that as late as 1893, when I had prepared an elaborate chapter on my wireless system, dwelling on its various instrumentalities and future prospects, Mr. Joseph Wetzler[4] and other friends of mine emphatically protested against its publication on the ground that such idle and far-fetched speculations would injure me in the opinion of conservative business men. So it came that only a small part of what I had intended to say was embodied in my address of that year before the Franklin Institute and National Electric Light Association under the chapter "On Electrical Resonance." This little salvage from the wreck has earned me the title of "Father of the Wireless" from many well-disposed fellow workers, rather than the invention of scores of appliances which have brought wireless transmission within the reach of every young amateur and which, in a time not distant, will lead to undertakings overshadowing in magnitude and importance all past achievements of the engineer.

The popular impression is that my wireless work was begun in 1893, but as a matter of fact I spent the two preceding years in investigations, employing forms of apparatus, some of which were almost like those of today. It was clear to me from the very start that the successful consummation could only be brought about by a number

---

[4] Joseph Wetzler (1863-1910) was an author prominently identified with electrical journalism, starting as a journalist with *Scientific American*, then serving as an editor for both *Electrical World* and *Electrical Engineer*. He authored several electrical books, and is credited as an editorial associate of T.C. Martin's *The Inventions, Researches and Writings of Nikola Tesla* (1894).

of radical improvements. Suitable high frequency generators and electrical oscillators had first to be produced. The energy of these had to be transformed in effective transmitters and collected at a distance in proper receivers. Such a system would be manifestly circumscribed in its usefulness if all extraneous interference were not prevented and exclusiveness secured. In time, however, I recognized that devices of this kind, to be most effective and efficient, should be designed with due regard to the physical properties of this planet and the electrical conditions obtaining on the same. I will briefly touch upon the salient advances as they were made in the gradual development of the system.

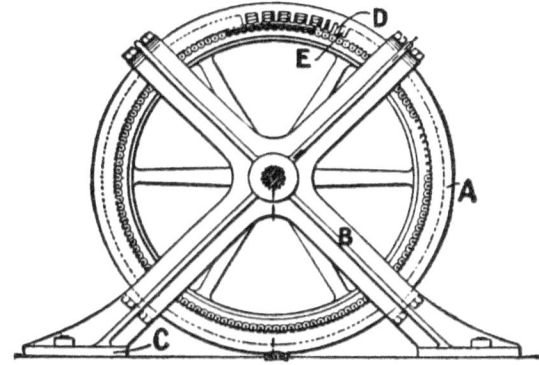

The high frequency alternator employed in my first demonstrations is illustrated in **Fig. 1**.

Fig. 1 — Alternator of 10,000 cycles per second, capacity 10 kilowatts, which was employed by Tesla in his first demonstrations of high frequency phenomena before the American Institute of Electrical Engineers at Columbia College, May 20, 1891.

It comprised a field ring, with 384 pole projections and a disc armature with coils wound in one single layer which were connected in various ways according to requirements. It was an excellent machine for experimental purposes, furnishing sinusoidal currents from 10,000 to 20,000 cycles per second. The output was comparatively large, due to the fact that as much as 30 amperes per square millimeter could be passed through the coils without injury.

The diagram in **Fig. 2** shows the circuit arrangements as used in my lecture. Resonant conditions were maintained by means of a condenser subdivided into small sections, the finer adjustments being effected by a movable iron core within an inductance coil. Loosely linked with the latter was a high tension secondary which was tuned to the primary.

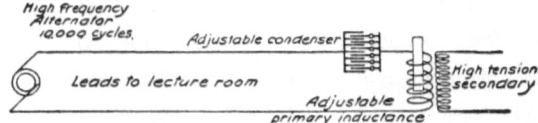

Fig. 2 — Diagram illustrating the circuit connections and tuning devices employed by Tesla in his experimental demonstrations before the American Institute of Electrical Engineers with the High Frequency Alternator shown in **Fig. 1**.

The operation of devices through a single wire without return was puzzling at first because of its novelty, but can be readily explained by suitable analogs. For this purpose reference is made to **Figs. 3 and 4**.

In the former the low resistance electrical conductors are represented by pipes of large section, the alternator by an oscillating piston and the filament of an incandescent lamp by a minute channel connecting the pipes. It will be clear from a glance at the diagram that very slight excursions of the piston would cause the fluid to rush with high velocity through the small channel and that virtually all the energy of movement would be transformed into heat by friction, similarly to that of the electric current in the lamp filament.

The second diagram will now be self-explanatory. Corresponding to the terminal capacity of the electric system an elastic reservoir is employed which dispenses

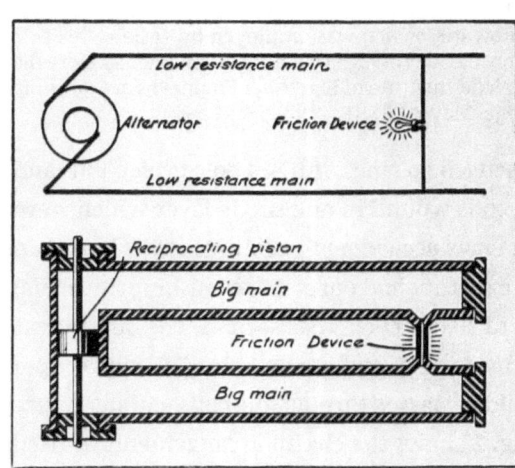

Fig. 3 — Electric transmission through two wires and hydraulic analog.

with the necessity of a return pipe. As the piston oscillates the bag expands and contracts, and the fluid is made to surge through the restricted passage with great speed, this resulting in the generation of heat as in the incandescent lamp. Theoretically considered, the efficiency of conversion of energy should be the same in both cases.

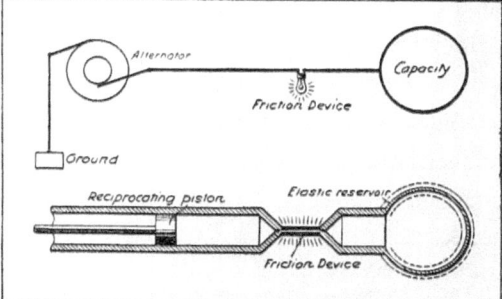

Fig. 4 — Electric transmission through a single wire hydraulic analog.

Granted, then, that an economic system of power transmission through a single wire is practicable, the question arises how to collect the energy in the receivers. With this object attention is called to **Fig. 5**, in which a conductor is shown excited by an oscillator joined to it at one end. Evidently, as the periodic impulses pass through the wire, differences of potential will be created along the same as well as at right angles to it in the surrounding medium and either of these may be usefully applied. Thus at *a*, a circuit comprising an inductance and capacity is resonantly excited in the transverse, and at *b*, in the longitudinal sense. At *c*, energy is collected in a circuit parallel to the conductor but not in contact with it, and again at *d*, in a circuit which is partly sunk into the conductor and may be, or not, electrically connected to the same. It is important to keep these typical dispositions in mind, for however the distant actions of the oscillator

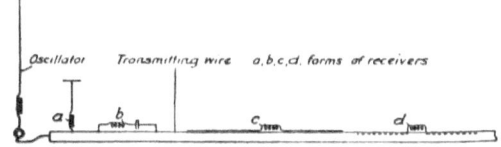

Fig. 5 — Illustrating typical arrangements for collecting energy in a system of transmission through a single wire.

might be modified through the immense extent of the globe the principles involved are the same.

Consider now the effect of such a conductor of vast dimensions on a circuit exciting it. The upper diagram of **Fig. 6** illustrates a familiar oscillating system comprising a straight rod of self-inductance $2L$ with small terminal capacities cc and a node in the center. In the lower diagram of the figure a large capacity C is attached to the rod at one end with the result of shifting the node to the right, through a distance corresponding to self-inductance X.

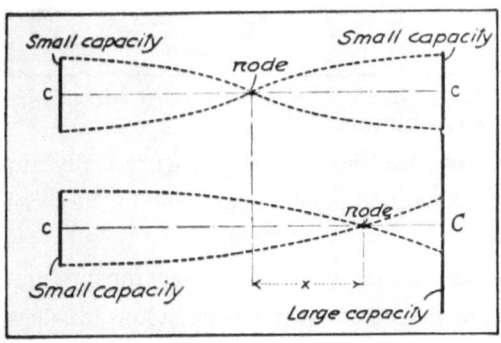

Fig. 6 — Diagram elucidating effect of large capacity on one end.

As both parts of the system on either side of the node vibrate at the same rate, we have evidently, $(L + X) c = (L - X) C$ from which $X = L (C - c / C + c)$. When the capacity C becomes commensurate to that of the earth, X approximates L, in other words, the node is close to the ground connection. *The exact determination of its position is very important in the calculation of certain terrestrial electrical and geodetic data* and I have devised special means with this purpose in view.

My original plan of transmitting energy without wires is shown in the upper diagram of **Fig. 7**, while the lower one illustrates its mechanical analog, first published in my article in *The Century Magazine* of June, 1900.[5] An alternator, preferably of high tension, has one of its terminals connected to the ground and the other to an elevated capacity and impresses its oscillations upon the earth. At a distant point a receiving circuit, likewise connected to ground and to an elevated

[5] Tesla, N. (1900, June). The Problem of Increasing Human Energy. *The Century Magazine*, 60(2), 175-211.

capacity, collects some of the energy and actuates a suitable device. I suggested a multiplication of such units in order to intensify the effects, an idea which may yet prove valuable. In the analog two tuning

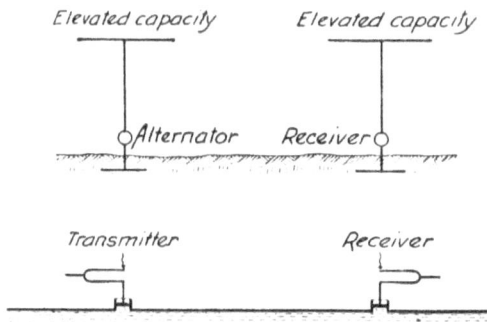

Fig. 7 — Transmission of electrical energy through the earth as illustrated in Tesla's lectures before the Franklin Institute and Electric Light Association in February and March 1893, and mechanical analog of the same.

forks are provided, one at the sending and the other at the receiving station, each having attached to its lower prong a piston fitting in a cylinder. The two cylinders communicate with a large elastic reservoir filled with an incompressible fluid. The vibrations transmitted to either of the tuning forks excite them by resonance and, through electrical contacts or otherwise, bring about the desired result. This, I may say, was not a mere mechanical illustration, but a simple representation of my apparatus for submarine signaling, perfected by me in 1892, but not appreciated at that time, although more efficient than the instruments now in use.

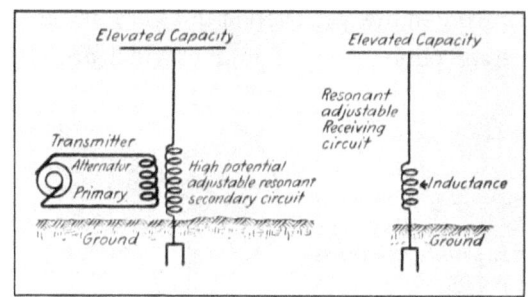

Fig. 8 — Tesla's system of wireless transmission through the earth as actually exposed in his lectures before the Franklin Institute and Electric Light Association in February and March 1893.

The electric diagram in **Fig. 7**, which was reproduced from my lecture, was meant only for the exposition of the principle. The arrangement, as I described it in detail, is shown in **Fig. 8**. In this case an alternator energizes the primary of a transformer, the high tension secondary of which is connected to the ground and an elevated capacity and tuned to the impressed oscillations. The receiving circuit consists of an inductance connected to the ground and to an elevated terminal without break and is resonantly responsive to the transmitted oscillations. A specific form of receiving device was not mentioned, but I had in mind to transform the received currents and thus make their volume and tension suitable for any purpose. This, in substance, is the system of today and I am not aware of a single authenticated instance of successful transmission at considerable distance by different instrumentalities. It might, perhaps, not be clear to those who have perused my first description of these improvements that, besides making known new and efficient types of apparatus, I gave to the world a wireless system of potentialities far beyond anything before conceived. I made explicit and repeated statements that I contemplated transmission, absolutely unlimited as to terrestrial distance and amount of energy. But, although I have overcome all obstacles which seemed in the beginning unsurmountable and found elegant solutions of all the problems which confronted me, yet, even at this very day, the majority of experts are still blind to the possibilities which are within easy attainment.

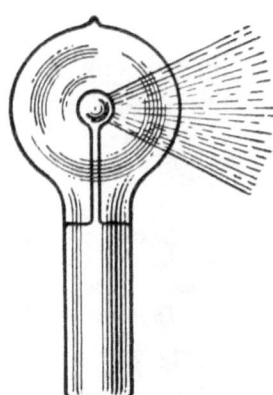

Fig. 9 — The forerunner of the audion — the most sensitive wireless detector known, as described by Tesla in his lecture before the Institution of Electrical Engineers, London, February 1892.

My confidence that a signal could be easily flashed around the globe was strengthened through the discovery of the "rotating brush," a wonderful phenomenon which I have fully described in my address before the Institution of Electrical Engineers, London, in 1892, and which is illustrated in **Fig. 9**. This is undoubtedly the most delicate wireless detector known, but for a long time it was hard to produce and to maintain in the sensitive state. These difficulties do not exist now and I am looking to valuable applications of this device, particularly in connection with the high-speed photographic method, which I suggested, in wireless, as well as in wire, transmission.

Possibly the most important advances during the following three or four years were my system of concatenated tuned circuits and methods of regulation, now universally adopted. The intimate bearing of these inventions on the development of the wireless art will appear from **Fig. 10**, which illustrates an arrangement described in

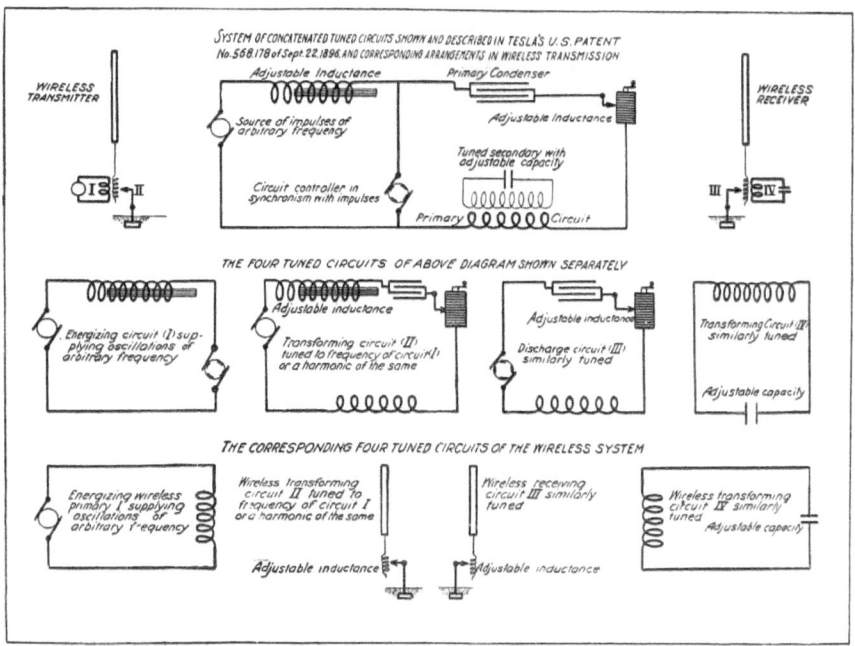

Fig. 10 — Tesla's system of concatenated tuned circuits shown and described in U.S Patent No. 568,178 of September 22, 1896, and corresponding arrangements in wireless transmission.

my U. S. Patent No. 568178 of September 22, 1896, and corresponding dispositions of wireless apparatus. The captions of the individual diagrams are thought sufficiently explicit to dispense with further comment. I will merely remark that in this early record, in addition to indicating how any number of resonant circuits may be linked and regulated, I have shown the advantage of the proper timing of primary impulses and use of harmonics. In a farcical wireless suit in London, some engineers, reckless of their reputation, have claimed that my circuits were not at all attuned; in fact they asserted that I had looked upon resonance as a sort of wild and untamable beast!

It will be of interest to compare my system as first described in a Belgian patent of 1897 with the Hertz-wave system of that period. The significant differences between them will be observed at a glance. The first enables us to transmit economically energy to any distance and is of inestimable value; the latter is capable of a radius of only a few miles and is worthless. In the first there are no spark-gaps and the actions are enormously magnified by resonance. In both transmitter and receiver the currents are transformed and rendered more effective and suitable for the operation of any desired device. Properly constructed, my system is safe against static and other interference and the amount of energy which may be transmitted is *billions of times greater* than with the Hertzian which has none of these virtues, has never been used successfully and of which no trace can be found at present.

A well-advertised expert gave out a statement in 1899 that my apparatus did not work and that it would take 200 years before a message would be flashed across the Atlantic and even accepted stolidly my congratulations on a supposed great feat. But subsequent examination of the records showed that my devices were secretly used all the time and ever since I learned of this I have treated these Borgia-Medici methods[6] with the contempt in which they are held by

---

[6] An allusion to the devilry the House of Borgia perpetrated against the House of Medici, particularly in the 15th and 16th centuries.

all fair-minded men. The wholesale appropriation of my inventions was, however, not always without a diverting side. As an example to the point I may mention my oscillation transformer operating with an air gap. This was in turn replaced by a carbon arc, quenched gap, an atmosphere of hydrogen, argon or helium, by a mechanical break with oppositely rotating members, a mercury interrupter or

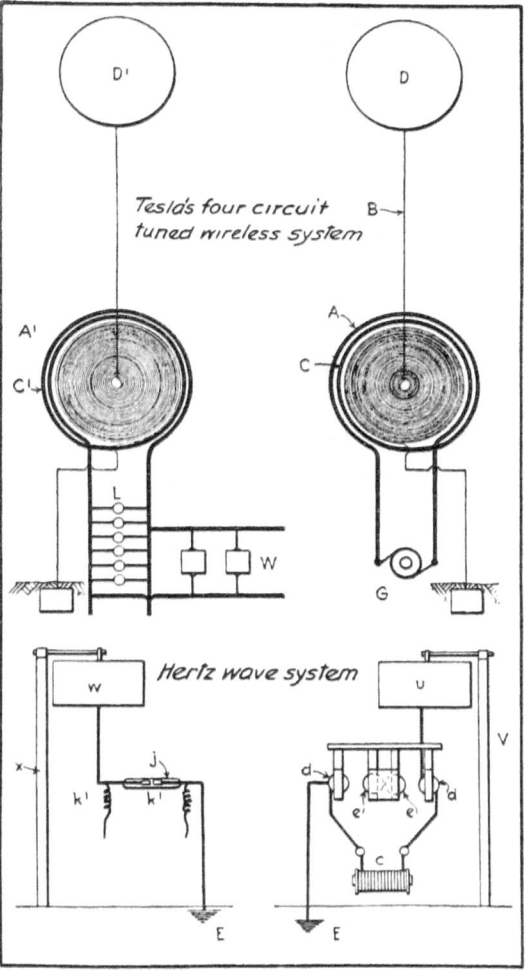

Fig. 11 — Tesla's four circuit tuned system contrasted with the contemporaneous Hertz-wave System.

some kind of a vacuum bulb and by such *tours de force* as many new "systems" have been produced. I refer to this of course, without the slightest ill-feeling, let us advance by all means. But I cannot help thinking how much better it would have been if the ingenious men, who have originated these "systems," had invented something of their own instead of depending on me altogether.

Before 1900 two most valuable improvements were made. One of these was my individualized system with transmitters emitting a wave-complex and receivers comprising separate tuned elements cooperatively associated. The underlying principle can be explained in a few words. Suppose that there are n simple vibrations suitable for use in wireless transmission, the probability that any one tune will be struck by an extraneous disturbance is $1/n$. There will then remain $n-1$ vibrations and the chance that one of these will be excited is $1/n-1$ hence the probability that two tunes would be struck at the same time is $1/n(n-1)$. Similarly, for a combination of three the chance will be $1/n(n-1)(n-2)$ and so on. It will be readily seen that in this manner any desired degree of safety against the statics or other kind of disturbance can be attained provided the receiving apparatus is so designed that its operation is possible only through the joint action of **all** the tuned elements. This was a difficult problem which I have successfully solved so that now *any desired number of simultaneous messages is practicable in the transmission through the earth as well as through artificial conductors.*

The other invention, of still greater importance, is a peculiar oscillator enabling the transmission of energy without wires in any quantity that may ever be required for industrial use, to any distance, and with very high economy. It was the outcome of years of systematic study and investigation and wonders will be achieved by its means.

The prevailing misconception of the mechanism involved in the wireless transmission has been responsible for various unwarranted announcements which have misled the public and worked harm.

By keeping steadily in mind that the transmission through the earth is in every respect identical to that through a straight wire, one will gain a clear understanding of the phenomena and will be able to judge correctly the merits of a new scheme. Without wishing to detract from the value of any plan that has been put forward I may say that they are devoid of novelty. So for instance in **Fig. 12** arrangements of transmitting and receiving circuits are illustrated, which I have described in my U. S. Patent No. 613809 of November 8, 1898, on a Method of and Apparatus for Controlling Mechanism of Moving Vessels or Vehicles, and which have been recently dished up as original discoveries. In other patents and technical publications I have suggested conductors in the ground as one of the obvious modifications indicated in **Fig. 5**.[7]

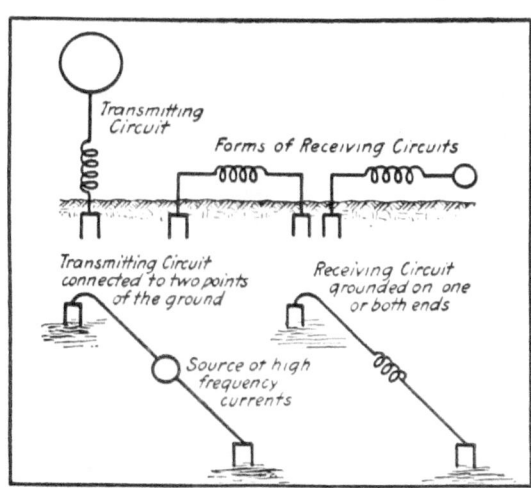

Fig. 12 — Arrangements of directive circuits described in Tesla's U.S. Patent No. 613,809 of November 8, 1898, on "Method of and Apparatus for Controlling Mechanism of Moving Vessels or Vehicles."

For the same reason the statics are still the bane of the wireless. There is about as much virtue in the remedies recently proposed as in hair restorers. *A small and compact apparatus has been produced which does away entirely with this trouble*, at least in plants suitably remodeled.

Nothing is more important in the present phase of development of the wireless art than to dispose of the dominating erroneous

[7] p. 145

ideas. With this object I shall advance a few arguments based on my own observations *which prove that Hertz waves have little to do with the results obtained even at small distances.*

In **Fig. 13** a transmitter is shown radiating space waves of considerable frequency. It is generally believed that these waves pass along the earth's surface and thus affect the receivers. I can hardly think of anything more improbable than this "gliding wave" theory and the conception of the "guided wireless" which are contrary to all laws of action and reaction. Why should these disturbances cling to a conductor where they are counteracted by induced currents, when they can propagate in all other directions unimpeded? The fact is that the radiations of the transmitter passing along the earth's surface are soon extinguished, the height of the inactive zone indicated in the diagram, being some function of the wave length,

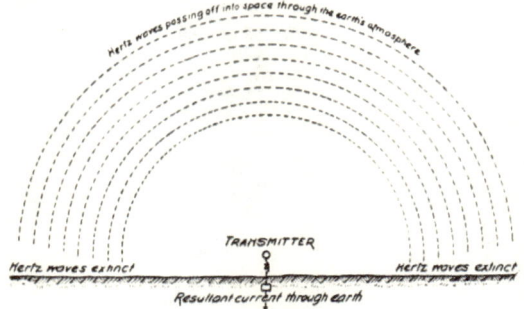

Fig. 13 — Diagram exposing the fallacy of the Gliding Wave Theory as propounded in wireless text books.

the bulk of the waves traversing freely the atmosphere. Terrestrial phenomena which I have noted conclusively show that there is no *Heaviside layer,* or if it exists, it is of no effect. It certainly would be unfortunate if the human race were thus imprisoned and forever without power to reach out into the depths of space.

The actions at a distance cannot be proportionate to the height of the antenna and the current in the same. I shall endeavor to make this clear by reference to the diagram in **Fig. 14**. The elevated terminal charged to a high potential induces an equal and opposite charge in the earth and there are thus Q lines giving an average current $I = 4Qn$ which circulates locally and is useless except that it adds to the momentum. A relatively small number of lines q however,

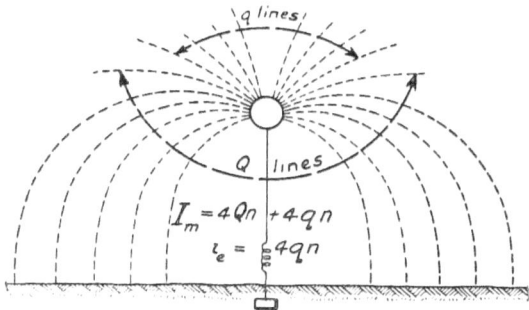

Fig. 14 — Diagram explaining the relation between the effective and the measured current in the antenna.

go off to great distance and to these corresponds a mean current of $ie = 4qn$ to *which is due the action at a distance*. The total average current in the antenna is thus $Im = 4Qn + 4qn$ and its intensity is no criterion for the performance. The electric efficiency of the antenna is $q / Q+q$ and this is often a very small fraction.

Dr. L. W. Austin[8] and Mr. J. L. Hogan[9] have made quantitative measurements which are valuable, but far from supporting the Hertz wave theory they are evidences in disproval of the same, as will be easily perceived by taking the above facts into consideration. Dr. Austin's researches are especially useful and instructive and I regret that I cannot agree with him on this subject. I do not think that if his receiver was affected by Hertz waves he could ever establish such relations as he has found, but he would be likely to reach these results if the Hertz waves were in a large part eliminated. At

---

[8] Louis Winslow Austin (1867–1932) was an American physicist known for his research on long-range radio transmissions.

[9] John Vincent Lawless Hogan (1890–1960) was a noted American radio pioneer.

great distance the space waves and the current waves are of equal energy, the former being merely an accompanying manifestation of the latter in accordance with the fundamental teachings of Maxwell.

It occurs to me here to ask the question — why have the Hertz waves been reduced from the original frequencies to those I have advocated for my system, when in so doing the activity of the transmitting apparatus has been reduced a billion fold? I can invite any expert to perform an experiment such as is illustrated in **Fig. 15**, which shows the classical Hertz oscillator and my grounded transmitting circuit. It is a fact which I have demonstrated that, although we may have in the Hertz oscillator an activity thousands of times greater, the effect on the receiver is not to be compared to that of the grounded circuit. This shows that *in the transmission from an airplane we are merely working through a condenser*, the capacity of which is a function of a logarithmic ratio between the length of the conductor and the distance from the ground. The receiver is affected in exactly the same manner as from an ordinary transmitter, the only difference being

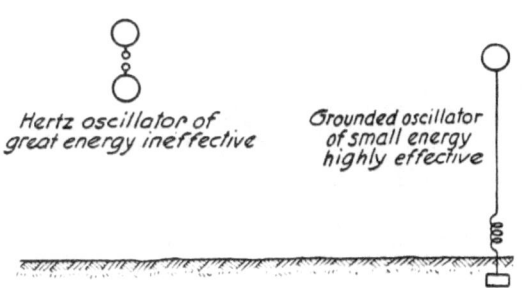

Fig. 15 — Illustrating one of the general evidences against the space wave transmission.

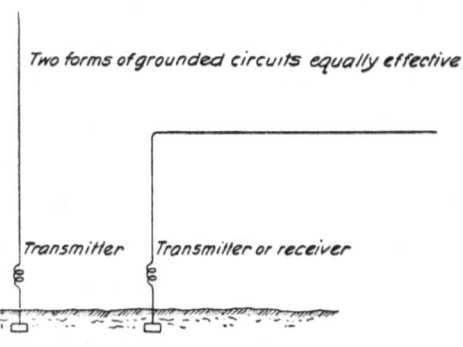

Fig. 16 — Showing unimportance of relative position of transmitting and receiving antennae in disproval of the Hertz-wave Theory.

that there is a certain modification of the action which can be predetermined from the electrical constants. It is not at all difficult to maintain communication between an airplane and a station on the ground, on the contrary, the feat is very easy.

To mention another experiment in support of my view, I may refer to **Fig. 16** in which two grounded circuits are shown excited by oscillations of the Hertzian order. It will be found that the

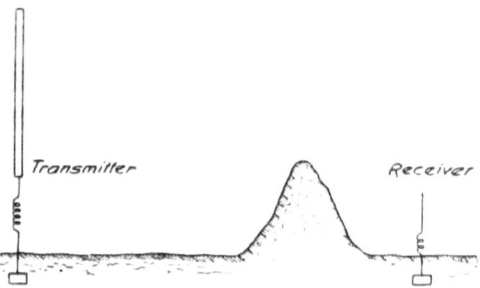

Fig. 17 — Illustrating influence of obstacles in the path of transmission as evidence against the Hertz-wave Theory.

antennas can be put out of parallelism without noticeable change in the action on the receiver, this proving that it is due to currents propagated through the ground and not to space waves.

Particularly significant are the results obtained in cases illustrated in **Figs. 17** and **18**. In the former an obstacle is shown in the path of the waves but unless the receiver is within the effective *electrostatic* influence of the mountain range, the signals are not appreciably weakened by the presence of the latter, because the currents pass under it and excite the circuit in the same way

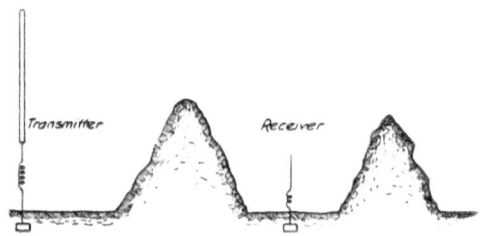

Fig. 18 — Showing effect of two hills as further proof against the Hertz-wave Theory.

as if it were attached to an energized wire. If, as in **Fig. 18**, a second range happens to be beyond the receiver, it could only strengthen the Hertz wave effect by reflection, but as a matter of fact it detracts greatly from the intensity of the received impulses because the electric niveau[10] between the mountains is raised, as I have explained with my lightning protector in the *Experimenter* of February.[11]

Again in **Fig. 19** two transmitting circuits, one grounded directly and the other through an air gap, are shown. It is a common observation that the former is far more effective, which could not be the case with Hertz radiations. In a like manner if two grounded circuits are observed from day to day the effect is found to increase greatly with the dampness of the ground, and for the same reason also the transmission through sea-water is more efficient.

An illuminating experiment is indicated in **Fig. 20** in which two grounded transmitters are shown, one with a large and the other with a small terminal capacity. Suppose that the latter be $\frac{1}{10}$ of the former but that it is charged to 10 times the potential and let the frequency of the two circuits and therefore the currents in both antennas be exactly the same. The circuit with the smaller capacity will then have 10 times the energy of the other but the effects on the receiver will be in no wise proportionate.

The same conclusions will be reached by transmitting and receiving circuits with wires buried underground. In each case the actions carefully investigated will be found to be due to *earth currents*. Numerous other proofs might be cited which can be easily verified. So for example *oscillations of low frequency* are ever so much more effective in the transmission which is inconsistent with the prevailing idea. My observations in 1900 and the recent transmissions of signals to very great distances are another emphatic disproval.

The Hertz wave theory of wireless transmission may be kept up for a while, but I do not hesitate to say that in a short time it will be recognized as one of the most remarkable and inexplicable aberrations of the scientific mind which has ever been recorded in history.

---

[10] A level or plateau, especially in a progression.
[11] p. 205, "Tesla Has New Pointless Lightning Rod"

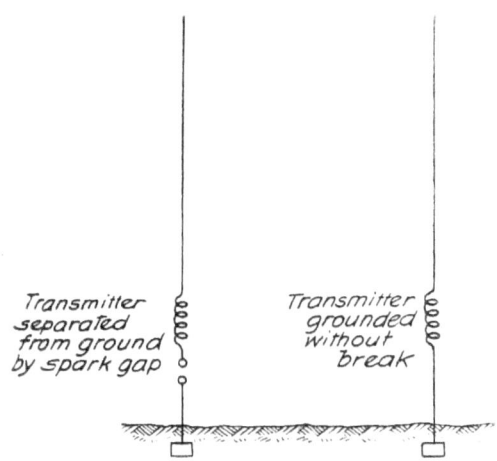

Fig. 19 — Comparing the actions of two forms of transmitter as bearing out the fallacy of the Hertz-wave Theory.

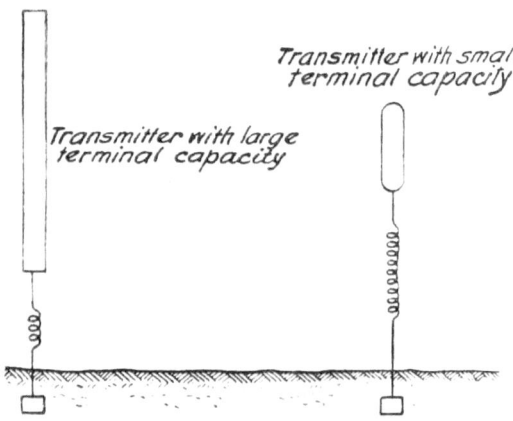

Fig. 20 — Disproving the Hertz-wave Theory by two transmitters, one of great and the other of small energy.

The Moon's Rotation

Since the appearance of my article entitled the "Famous Scientific Illusions"[1] in your February issue, I have received a number of letters criticizing the views I expressed regarding the moon's "axial rotation." These have been partly answered by my statement to the *New-York Tribune* of February 23, 1919, which allow me to quote:

In your issue of February 2, Mr. Charles E. Manierre, commenting upon my article in the *Electrical Experimenter* for February which appeared in the *Tribune* of January 26, suggests that I give a definition of axial rotation.

I intended to be explicit on this point as may be judged from the following quotation: "The unfailing test of the spinning of a mass is, however, the existence of *energy of motion*. The moon is not possessed of such *vis viva*." By this I meant that "axial rotation" is not simply "rotation upon an axis nonchalantly defined in dictionaries, but is a circular motion in the true physical sense — that is, one in which half the product of the mass with the square of velocity is a definite and positive quantity. The moon is a nearly spherical body, of a radius of about 1,087.5 miles, from which I calculate its volume to be approximately 5,300,216,300 cubic miles. Since its

[1] p. 113

mean density is 3.27, one cubic foot of material composing it weighs close on 205 lbs. Accordingly, the total weight of the satellite is about 79,969,000,000,000,000,000, and its mass 2,483,500,000,000,000,000 terrestrial short tons. Assuming that the moon does physically rotate upon its axis, it performs one revolution in 27 days, 7 hours, 43 minutes and 11 seconds, or 2,360,591 seconds. If, in conformity with mathematical principles, we imagine the entire mass concentrated at a distance from the center equal to two-fifths of the radius, then the calculated rotational velocity is 3.04 feet per second, at which the globe would contain 11,474,000,000,000,000,000 short foot tons of energy sufficient to run 1,000,000,000 horsepower for a period of 1,323 years. Now, I say, that there is not enough of that energy in the moon to run a delicate watch.

In astronomical treatises usually the argument is advanced that "if the lunar globe did not turn upon its axis it would expose all parts to terrestrial view. As only a little over one-half is visible it *must* rotate." But this inference is erroneous, for it only admits of one alternative. There are an infinite number of axis besides its own in each of which the moon might turn and still exhibit the same peculiarity.

I have stated in my article that the moon rotates about an axis passing through the center of the earth, which is not strictly true, but it does not vitiate[2] the conclusions I have drawn. It is well known, of course, that the two bodies revolve around a common center of gravity, which is at a distance of a little over 2,899 miles from the earth's center. Another mistake in books on astronomy is made in considering this motion equivalent to that of a weight whirled on a string or in a sling. In the first place there is an essential difference between these two devices though involving the same mechanical principle. If a metal ball, attached to a string, is whirled around and the latter breaks, an axial rotation of the missile results which

---

[2] Spoil or impair the quality or efficiency of.

is definitely related in magnitude and direction to the motion preceding. By way of illustration — if the ball is whirled on the string clockwise ten times per second, then when it flies off, it will rotate on its axis ten times per second, likewise in the direction of a clock. Quite different are the conditions when the ball is thrown from a sling. In this case a *much more rapid* rotation is imparted to it in the *opposite sense*. There is no true analogy to these in the motion of the moon. *If the gravitational string, as it were, would snap, the satellite would go off in a tangent without the slightest swerving or rotation, for there is no moment about the axis and, consequently, no tendency whatever to spinning motion.*

Mr. Manierre is mistaken in his surmise as to what would happen if the earth were suddenly eliminated. Let us suppose that this would occur at the instant when the moon is in *opposition*. Then it would continue on its elliptical path around the sun, presenting to it steadily the face which was always exposed to the earth. If, on the other hand, the latter would disappear at the moment of *conjunction*, the moon would gradually swing around through 180° and, after a number of oscillations, revolve, again with the same face to the sun. In either case there would be no periodic changes but eternal day and night, respectively, on the sides turned toward, and away from, the luminary. Some of the arguments advanced by the correspondents are ingenious and not a few comical. None, however, are valid.

One of the writers imagines the earth in the center of a circular orbital plate, having fixedly attached to its peripheral portion a disk-shaped moon, in frictional or geared engagement with another disk of the same diameter and freely rotatable on a pivot projecting from an arm entirely independent of the planetary system. The arm being held continuously parallel to itself, the pivoted disk, of course, is made to turn on its axis as the orbital plate is rotated. This is a well-known drive, and the rotation of the pivoted disk is

164   **MY INVENTIONS** & other essays

as palpable a fact as that of the orbital plate. But, the moon in this model only revolves about the center of the system *without the slightest angular displacement* on its own axis. The same is true of a cart-wheel to which this writer refers. So long as it advances on the earth's surface it turns on the axle in the true physical sense; when one of its spokes is always kept in a perpendicular position the wheel still *revolves* about the earth's center, *but axial rotation has ceased*. Those who think that it then still exists are laboring under an illusion.

An obvious fallacy is involved in the following abstract reasoning. The orbital plate is assumed to gradually shrink, so that finally the centers of the earth and the satellite coincide when the latter revolves simultaneously about its own and the earth's axis. We may reduce the earth to a mathematical point and the distance between the two planets to the radius of the moon without affecting the system in principle, but a further diminution of the distance is manifestly absurd and of no bearing on the question under consideration.

In all the communications I have received, though different in the manner of presentation, the successive changes of position in space are mistaken for axial rotation. So, for instance, a positive refutation of my arguments is found in the observation that the moon exposes all sides to other planets! It revolves, to be sure, but none of the evidences is a proof that it turns on its axis. Even the well-known experiment with the Foucault pendulum,[3] although exhibiting similar phenomena as on our globe, would merely demonstrate a motion of the satellite about *some* axis. The view I have advanced is NOT BASED ON A THEORY but on facts *demonstrable by experiment*. It is not a matter of *definition* as some would have it. A MASS REVOLVING ON ITS AXIS MUST BE POSSESSED OF

---

[3] A device created by French physicist Léon Foucault (1819–1868) in 1851 and conceived as an experiment to demonstrate simple, direct evidence of the Earth's rotation.

MOMENTUM. If it has none, there is no axial rotation, all appearances to the contrary notwithstanding.

A few simple reflections based on well established mechanical principles will make this clear. Consider first the case of two equal weights w and w_1, in Fig. 1, whirled about the center O on a string s as shown. Assuming the latter to break at a both weights will fly off on tangents to their circles of gyration, and, being animated with different velocities, they

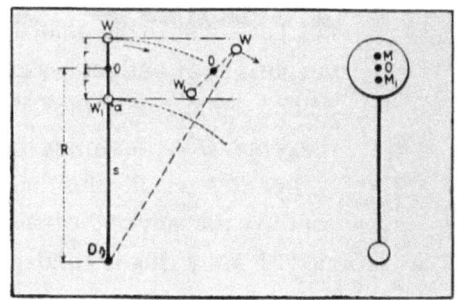

Fig. 1 — Diagram illustrating the rotation of weights thrown off by centrifugal force.

will rotate around their common center of gravity o. If the weights are whirled n times per second then the speed of the outer and the inner one will be, respectively, $V = 2(R + r)n$ and $V_1 = 2\pi(R-r)n$, and the difference $V - V_1 = 4\pi r n$, will be the length of the circular path of the outer weight. Inasmuch, however, as there will be equalization of the speeds until the mean value is attained, we shall have, $V - V_1/2 = 2\pi r n = 2\pi r N$, N being the number of revolutions per second of the weights around their center of gravity. Evidently then, the weights continue to rotate at the original rate and in the same direction. I know this to be a fact from actual experiments. It also follows that a ball, as that shown in the figure, will behave in a similar manner for the two half-spherical masses can be concentrated at their centers of gravity and m and m_1, respectively, which will be at a distance from o equal to $\frac{3}{8}r$.

This being understood, imagine a number of balls M carried by as many spokes S radiating from a hub H, as illustrated in Fig. 2, and let this system be rotated n times per second around center O on frictionless bearings. A certain amount of work will be required to bring the structure to this speed, and it will be found that it

We believe the accompanying illustration and its explanation will dispel all doubts as to whether the moon rotates on its axis or not. Each of the balls, as *M*, depicts a different position of, and rotates exactly like, the moon keeping always the same face toward the center *O*, representing the earth.

But as you study this diagram, can you conceive that any of the balls turn on their axis? Plainly this is rendered physically impossible by the spokes. But if you are still unconvinced, Mr. Tesla's experimental proof will surely satisfy you. A body rotating on its axis must contain rotational energy. Now it is a fact, as Mr. Tesla shows, that no such energy is imparted to the ball as, for instance, to a projectile discharged from a gun. It is therefore evident that the moon, in which the gravitational attraction is substituted for a spoke, cannot rotate on its axis or, in other words, contain rotational energy. If the earth's attraction would suddenly cease and cause it to fly off in a tangent, the moon would have no other energy except that of translatory movement, and it would not spin like the ball.

—Editor.
Electrical Experimenter
April 1919

equals exactly half the product of the masses with the square of the tangential velocity. Now if it be true that the moon rotates in reality on its axis *this must also hold good for EACH of the balls as it performs the same kind of movement*. Therefore, in imparting to the system a given velocity, energy must have been used up in the axial rotation of the balls. Let *M* be the mass of one of these and *R* the radius of gyration, then the rotational energy will be *E = ½M (2 π R n)²*. Since for one complete turn of the wheel every ball makes one revolution on its axis, according to the prevailing theory, the energy of axial rotation of each ball will be *e = ½M (2 π r₁ n)²*, *r₁* being the radius of gyration about the axis and equal to 0.6325 *r*. We can use balls as large as we like, and so make *e* a considerable percentage of *E* and yet, it is positively established by experiment that each of the rotating balls contain only the energy *E*, no power whatever being consumed in the supposed axial rotation, which is,

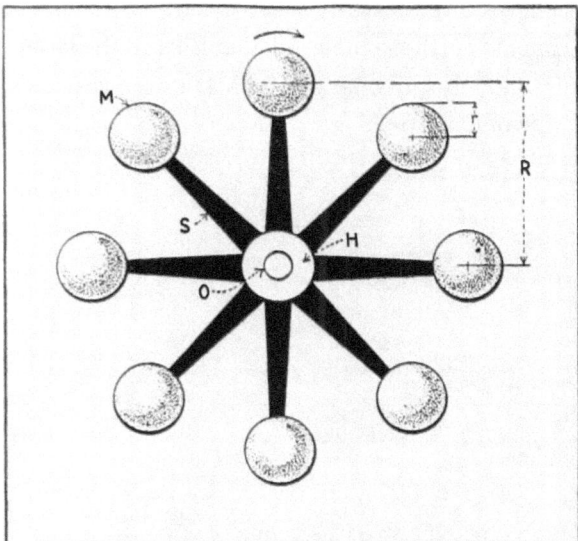

Fig. 2 — If you still think that the moon rotates on its axis, look at this diagram and follow closely the successive positions taken by one of the balls, *M*, while it is rotated by a spoke and the analogy solves the moon rotation riddle.

consequently, wholly illusionary. Something even more interesting may, however, be stated. As I have shown before, a ball flying off will rotate at the rate of the wheel and in the same direction. But this whirling motion, unlike that of a projectile, neither adds to, nor detracts from, the energy of the translatory movement which is exactly equal to the work consumed in giving to the mass the observed velocity.

From the foregoing it will be seen that in order to make one physical revolution on its axis the moon should have twice its present angular velocity, and then it would contain a quantity of stored energy as given in my above letter to the *New-York Tribune*, on the assumption that the radius of gyration is ⅔ that of figure. This, of course, is uncertain, as the distribution of density in the interior is unknown. But from the character of motion of the satellite it may be concluded with certitude *that it is devoid of momentum about its axis*. If it be bisected by a plane tangential to the orbit, the masses of the two halves are inversely as the distances of their centers of gravity from the earth's center and, therefore, if the latter were to disappear suddenly, no axial rotation, as in the case of a weight thrown off, would ensue.

In this article Dr. Tesla proves conclusively by theory and experiment that all the kinetic energy of a rotating mass is purely translational and that the moon contains absolutely no rotational energy, in other words, does not rotate on its axis.

—Editor.
Electrical Experimenter
June 1919

The Moon's Rotation (follow-up)

In revising my article on "The Moon's Rotation," which appeared in the April issue of the *Electrical Experimenter*, I appended a few remarks to the original text in further support and elucidation of the theory advanced. Due to the printer's error these were lost and, in consequence, I found it necessary to forward another communication which, unfortunately, was received too late for embodiment in the May number. Meanwhile many letters have reached me in which certain phenomena presented by rotating bodies, as the moon's librations of longitude, are cited as evidences of energy due to spinning motion, i.e., proofs of axial rotation of the satellite in the true physical sense. I trust that the following amplified statement will meet all of the objections raised and convert to my views those who are still unconvinced.

The kinetic energy of a rotating mass can be determined in four ways which are illustrated in **Figs. 1, 2, 3**, and **4** and may be found more or less suitable.

Referring to **Fig. 1**, the method consists in selecting judiciously a number of points as O_1, O_2, O_3, etc., within the straight rod or mass

M, respectively at distances r_1, r_2, r_3, etc., from the axis of rotation O and calculating the square root of the mean square of these distances. Its value being R_g, denoted radius of gyration, the *effective* velocity of the mass at n revolutions per second will be $V_e = 2\pi R_g n$ and its kinetic energy $E = \frac{1}{2} M V_e^2 = \frac{1}{2} M (2\pi R_g n)^2$.

In **Fig.** 2 the mass M, rotating n times per second about an axis O at right angles to the plane of the paper, is divided into numerous elements or small parts, most conveniently very thin concentric laminae, as 1_1, 1_2, 1_3, etc., at distances r_1, r_2, r_3, etc., from O. Since the kinetic energy of each part is equal to half the product of its mass and the square of the velocity, the sum of all these elemental energies $E = \frac{1}{2}\Sigma m V^2 = \frac{1}{2} m_1 V_1^2 + \frac{1}{2} m_2 V_2^2 + \frac{1}{2} m_3 V_3^2 + \ldots = \frac{1}{2} m_1 (2\pi r_1 n)^2 + \frac{1}{2} m_2 (2\pi r_2 n)^2 + \frac{1}{2} m_3 (2\pi r_3 n)^2 + \ldots$.

A different form of expression for the energy of a rotating body may be obtained by determining its moment of inertia. For this purpose the mass M (in **Fig. 3**), rotating n times per second about an axis O, is separated into minute parts, as m_1, m_2, m_3, etc., respectively at distances r_1, r_2, r_3, etc., from the same. The sum of the products of all these small masses and the squares of their distances is the moment of inertia I, and then $E = \frac{1}{2} I \omega^2$, $\omega = 2\pi n$ being the angular velocity.

It is obvious that in all these instances many points or elements will be required for great accuracy but, as a rule, very few are sufficient in practice.

Still another way to compute the kinetic energy is illustrated in **Fig. 4**, in which case the quantity I is given in terms of the moment of inertia I_e about another axis parallel to O and passing through the center of gravity C of mass M. In conformity with this the energy of motion $E = \frac{1}{2} M V^2 + \frac{1}{2} I_e \omega^2$ in which equation V is the velocity of the center of gravity.

The preceding is deemed indispensable as I note that the correspondents, even those who seem thoroughly familiar with mechanical

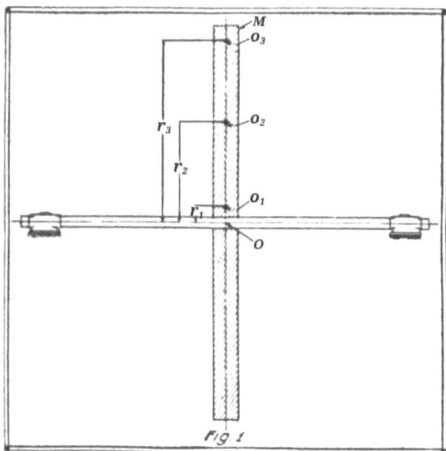

Fig. 1 — In determining the kinetic energy of the rotating mass, this figure shows the selection of a number of points taken within the straight rod or mass *M*, at successive distances from the axis of rotation *O*, knowing these values and the speed of rotation the kinetic energy of the mass is readily computed.

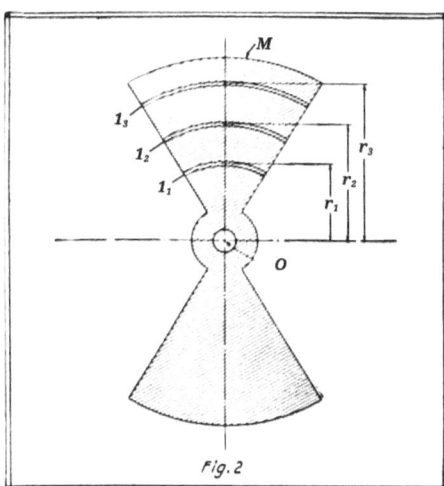

Fig. 2 — In this case the mass *M*, rotating *n* times per second, about an axis *O*, is divided into numerous elements or small parts at various radii from *O*. Knowing the kinetic energy of each part, the whole kinetic energy of the mass is easily determined by taking a summation of the individual quantities.

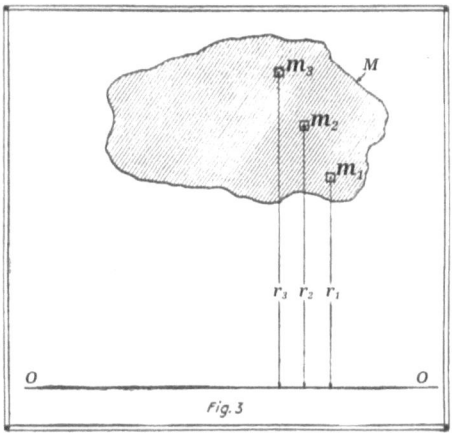

Fig. 3 — Another form of expression for the energy of a rotating body may be obtained by determining its moment of inertia. Here the mass *M* is subdivided into minute parts m_1, m_2, m_3, ... etc. The sum of the products of these masses and the squares of their distances is the moment of inertia, which with the angular speed, gives the kinetic energy *E*.

principles, fail to make a distinction between theoretical and physical truths which is essential to my argument.

In estimating the kinetic energy of a rotating mass in any of the ways indicated we arrive, through suitable conceptions and methods of approximation, at expressions which may be made quantitatively precise to any desired degree, but do not truly define the actual condition of the body. To illustrate, when proceeding according to the plan of **Fig. 1**, we find a certain hypothetical velocity with which the entire mass should move in order to contain the same energy, a state wholly imaginary and irreconcilable with the actual. Only, when *all* particles of the body have the same velocity, does the product ½ $M V^2$ specify a physical fact and is numerically and descriptively accurate. Still more remote from palpable truth is the equation of motion obtained in the manner indicated in **Fig. 4**, in which the first term represents the kinetic energy of translation of the body as a whole and the second that of its axial rotation. The former would demand a movement of the mass in a definite path and direction, all particles having the same velocity, the latter its simultaneous motion in another path and direction, the particles having different velocities. This abstract idea of angular motion is chiefly responsible for the illusion of the moon's axial rotation, which I shall endeavor to dispel by additional evidences.

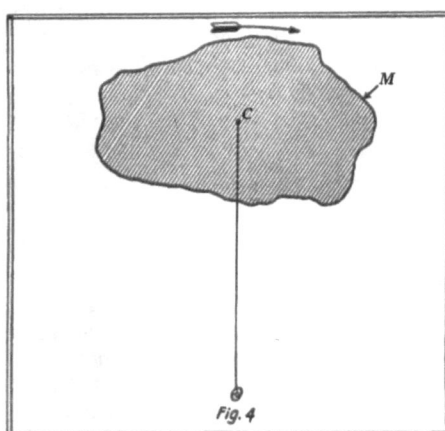

Fig. 4 — In this case the motion is resolved into two separate components — one translational about *O* and the other rotational about *C*. The total kinetic energy of the mass equals the sum of these two energies.

The Moon's Rotation (follow-up)

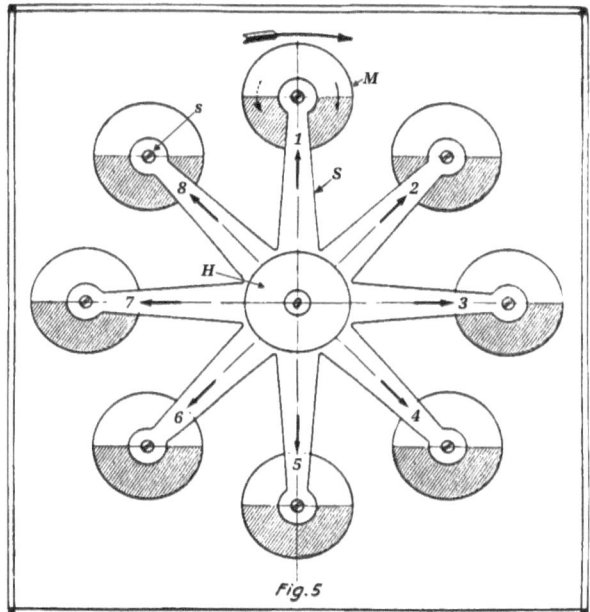

Fig. 5 — This diagram represents a system composed of 8 balls *M*, carried on spokes *S*, and rotating around center *O*. The balls are freely rotatable on pivots *s* which can be tightened. With this model the fallacy of the moon's rotation on its axis is demonstrable.

With this object attention is called to **Fig. 5** showing a system composed of eight balls *M*, which are carried on spokes *S*, radiating from a hub *H*, rotatable around a central axis *O* in bearings supposed to be frictionless. It is an arrangement similar to that before illustrated with the exception that the balls, instead of forming parts of the spokes, are supported in screw pivots *s*, which are normally loose but can be tightened so as to permit both free turning and rigid fixing as may be desired. To facilitate observation the spokes are provided with radial marks and the lower sides of the balls are shaded. Assume, first, that the drawing depicts the state of rest, the balls being rotatable without friction, and let an angular velocity $\omega = 2\pi n$ be imparted to the system in the clockwise direction as indicated by the long solid arrow. Viewing a ball as M, its successive positions 1, 2, 3–8 in space, and also relatively to the spoke, will be just as drawn, and it is evident from an inspection of the diagram that while moving with the angular velocity ω about O, in the clockwise direction, the ball turns, with respect to its axis, at the same angular velocity but in the opposition direction, that of the dotted arrow. The combined result of these two motions is a translatory movement of the ball such that all particles are animated with the same velocity V, which is that of its center of gravity. In this case, granted that there is absolutely no friction the kinetic energy of each ball will be given by the product of $½ M V^2$ not approximately, but with mathematical rigor. If now the pivots are screwed tight and the balls fixed rigidly to the spokes, this angular motion relatively to their axes becomes physically impossible and then it is found that the kinetic energy of each ball is increased, the increment being exactly the energy of rotation of the ball on its axis. This fact, which is borne out both by theory and experiment, is the foundation of the general notion that a gyrating body — in this instance ball M — presenting always the same face toward the center of motion, *actually* rotates upon its axis in the same sense, as indicated by the short full arrow. *But it does not though to the eye it seems so.* The fallacy will become manifest on further inquiry.

To begin with, observe that when a mass, say the armature of an electric motor, rotating with the angular velocity ω, is reversed, its speed is $-\omega$ and the difference $\omega - (-\omega) = 2\omega$. Now, in fixing the ball to the spoke, the change of angular velocity is only ω; therefore, an additional velocity ω would have to be imparted to it in order to cause a clockwise rotation of the ball on its axis in the true significance of the word. The kinetic energy would then be equal to the sum of the energies of the translatory and axial motions, not merely in the abstract mathematical meaning, but as a physical fact. I am well aware that, according to the prevailing opinion, when the ball

The Moon's Rotation (follow-up)

is free on the pivots it does not turn on its axis at all and only rotates with the angular velocity of the frame when rigidly attached to the same, but the truth will appear upon a closer examination of this kind of movement.

Let the system be rotated as first assumed and illustrated, the balls being perfectly free on the pivots, and imagine the latter to be gradually tightened to cause friction slowly reducing and finally preventing the slip. At the outset all particles of each ball have been moving with the speed of its center of gravity, but as the bearing resistance asserts itself more and more the translatory velocity of the particles nearer to the axis O will be *diminishing*, while that of the diametrically opposite ones will be *increasing*, until the maxima of these changes are attained when the balls are firmly held. In this operation we have thus deprived those parts of the masses which are nearer to the center of motion, of some kinetic energy of *translation* while adding to the energy of those which are farther and, obviously, the gain was greater than the loss so that the *effective* velocity of each ball as a whole was increased. *Only so* have we augmented the kinetic energy of the system, not by causing *axial rotation* of the balls. The energy E of each of these is *solely* that of translatory movement with an effective velocity V_e as above defined such that $E = \frac{1}{2} M V_e^2$. The axial rotations of the ball in either direction are but apparent; *they have no reality whatever* and call for no mechanical effort. It is merely when an extraneous force acts independently to turn the whirling body on its axis that energy comes into play. Incidentally it should be pointed out that in true axial rotation of a rigid and homogenous mass all symmetrically situated particles contribute equally to the momentum which is not the case here. That there exists not even the slightest tendency to such motion can, however, be readily established.

For this purpose I would refer to **Fig. 6** showing a ball M of radius r, the center C of which is at a distance R from axis O and which is bisected by a tangential plane pp as indicated, the lower half sphere being shaded for distinction. The kinetic energy of the ball when whirled n times per second about O is according to the first form of expression $E = \frac{1}{2} M V_e^2 = \frac{1}{2} M (2 \pi R_g n)^2$, M being the mass and R_g the radius of gyration. But, as explained in connection with **Fig. 4**, we have also $E = \frac{1}{2} M V^2 + \frac{1}{2} I_e \omega^2$, $V = 2 \pi R n$ being the velocity of the center of gravity C and I_e the moment of inertia of the ball, about the parallel axis passing through C and equal to $\frac{2}{5} M i^2$ so that $E = \frac{1}{2} M (2 \pi R n)^2 + \frac{1}{5} M r^2 (2 \pi n)^2$. Neither of these two expressions for E describes the actual state of the body but the first is certainly preferable conveying, as it does, the idea of a single motion instead of two, one of which moreover is devoid of existence. I shall first undertake to demonstrate that there is no torque or rotary effort about center C and that the kinetic energy of the supposed axial rotation of the ball is mathematically equal to zero. This makes it necessary to consider the two halves separated by the tangential plane pp wholly independent from one another. Let c_1 and c_2 be their centers of gravity, then $Cc_1 = Cc_2 = \frac{3}{8} r$. In order to ascertain the kinetic energy of the hemispheres we have to find their radii of gyration which can be done by determining the moments of inertia

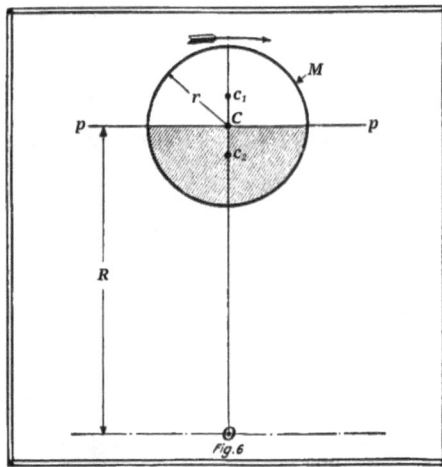

Fig. 6 — Diagram showing a ball having mass M, of radius r, rotating about center O, and used in theoretical analysis of the moon's motion.

The Moon's Rotation (follow-up)

Ic_1 and Ic_2 about parallel axes passing through c_1 and c_2. Complex calculation will be avoided by remembering that the moment of inertia of either one of the half spheres about an axis thru C is $Ic = \frac{1}{2} \times \frac{2}{5} M r^2 = \frac{1}{5} M r^2$, and since $M = 2m$, $Ic = \frac{2}{5} m r^2$. This can be expressed in terms of the moments Ic_1 and Ic_2; namely, $Ic = Ic_1 + m (\frac{3}{8} r)^2 = Ic_2 + m (\frac{3}{8} r)^2$. Hence $Ic_1 = Ic_2 = Ic - m (\frac{3}{8} r)^2 = \frac{2}{5} m r^2 - 9/64 \, m r^2 = 83/320 \, m r^2$. Following the same rule the moments of inertia of the half spheres about the axis passing through the center of motion O can be found. Designating the moments for the upper and lower halves of the ball, respectively, Io_1 and Io_2 we have $Io_1 = m (R + \frac{3}{8} r)^2 + Ic_1 = m (R + \frac{3}{8} r)^2 + 83/320 \, m r^2$ and $Io_2 = m (R - \frac{3}{8} r)^2 + Ic_2 = m (R - \frac{3}{8} r)^2 + 83/320 \, m r^2$. Thus for the upper half sphere the radius of gyration $R_{g1} = \sqrt{Io_1 / m} = \sqrt{(R + \frac{3}{8} r)^2 + 83/320 \, r^2}$ and for the lower one $R_{g2} = \sqrt{Io_2 / m} = \sqrt{(R - \frac{3}{8} r)^2 + 83/320 \, r^2}$.

These are the distances from center O, at which the masses of the half spheres may be concentrated and then the algebraic sum of their energies — which are wholly translatory those of axial rotation being nil — will be exactly equal to the total kinetic energy of the ball as a unit. The significance of this will be understood by reference to **Fig. 7** in which the two masses, condensed into points, are represented as attached to independent weightless strings of lengths R_{g1} and R_{g2}

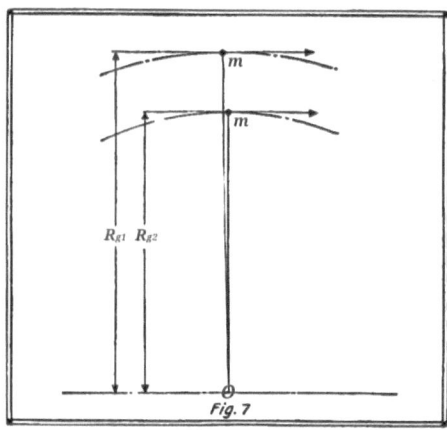

Fig. 7

Fig. 7 — Here two masses *m-m* are considered as condensed into points, attached to weightless strings of different radii. If both string are cut, and the masses considered joined, then there will be no rotation about the common center of gravity.

ch. ends p. 184

which are purposely shown as displaced but should be imagined as coincident. It will be readily seen that if both strings are cut in the same instant the masses will fly off in tangents to their circular orbits, the angular movement becoming rectilinear without any transformation of energy occurring. Let us now inquire what will happen if the two masses are rigidly joined, the connection being assumed imponderable. *Here we come to the real bug in the question under discussion.* Evidently, so long as the whirling motion continues, and both the masses have precisely the same angular velocity, this connecting link will be of no effect whatever, not the slightest turning effort about the common center of gravity of the masses or tendency of equalization of energy between them will exist. The moment the strings are broken and they are thrown off they will begin to rotate but, as pointed out before, this motion neither adds to or detracts from the energy stored. **The rotation is, however, not due to an exclusive virtue of angular motion, but to the fact that the tangential velocities of the masses or parts of the body thrown off are different.**

To make this clear and to investigate the effects produced, imagine two rifle barrels, as shown in **Fig. 8**, placed parallel to each other with their axes separated by a distance $R_{g1} - R_{g2}$ and assume that two balls of same diameter, each having mass m, are discharged with muzzle velocities V_1 and V_2, respectively equal to $2 \pi n R_{g1}$ and $2 \pi n R_{g2}$ as in the case just considered. If it be further supposed that at the instant of leaving the barrels the balls are joined by a rigid but weightless link they will rotate about their common center of gravity and in accordance with the statement in my previous article above mentioned, the relation will exist $V_1 - V_2 / 2 = \pi n (R_{g1} - R_{g2})$, n being the number of revolutions per second. The equalization of the speeds and kinetic energies of the balls will be, under these circumstances, very rapid but in two heavenly bodies linked by gravitational attraction, the process might require ages. Now, this

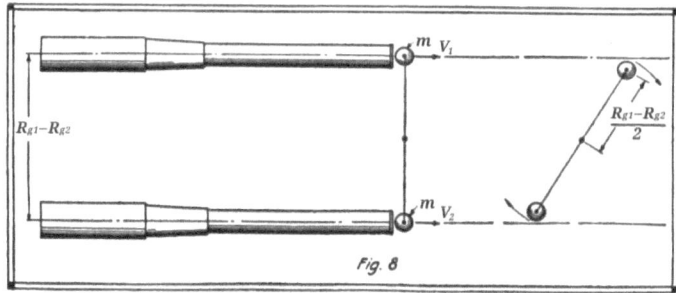

Fig. 8 — To make the problem shown in **Fig. 7** clear, imagine two rifle barrels parallel to each other. If two balls *m-m* are fired simultaneously, joined by a theoretical bond, they will revolve about their common center of gravity, proving that the moon possesses only kinetic energy of translation.

whirling movement is real and requires energy which, obviously, must be derived from that originally imparted and, consequently, must reduce the velocity of the balls in the direction of flight by an amount which can be easily calculated. At the moment of discharge the total kinetic energy was $E = \frac{1}{2} m V_1^2 + \frac{1}{2} m V_2^2$ which is evidently equal to $m V_3^2$, V_3 being the effective velocity of the common center of gravity, from which follows that $V_3 = \sqrt{V_1 + V_2 / 2}$. The speed of revolution of the masses is, of course, $V_1 - V_2 / 2$ and the rotational energy of both balls, which must be considered as points, is $e = m (V_1 - V_2 / 2)$. The kinetic energy of translation in the direction of flight is then $\frac{1}{2} m V_1^2 + \frac{1}{2} m V_2^2 - m (V_1 - V_2 / 2)^2 = m (V_1 + V_2 / 2)^2 = m V_4^2$, $V_4 = V_1 + V_2 / 2$ being the speed of the common center of gravity, so that $V_3 - V_4$ is the loss of velocity in the direction of flight owing to the rotation of the two mass points. If instead of these we would deal with the balls as they are, their rotational energy $e_1 = e + i \omega^2 = m (V_1 + V_2 / 2)^2 + i (2 \pi n)^2$, *i* being the moment of inertia of each ball about its axis.

As will be seen, we arrive at precisely the same results whether the movement is rectilinear or in a circle. In both cases the total kinetic energy can be divided into two parts, respectively of the same numerical values, *but there is an essential difference*. In angular

ch. ends p. 184

motion the axial rotation is nothing more than an *abstract conception*; in rectilinear movement it is a *positive event*.

Virtually all satellites rotate in like manner and the probability that the acceleration or retardation of their axial motions — if they ever existed — should come to a stop precisely at a definite angular velocity is infinitesimal while it is almost absolutely certain that all movement of this kind would ultimately cease. The most plausible view is that no true moon has ever rotated on its axis, for at the time of its birth there must have been some deformation and displacement of its center of gravity through the attractive force of the mother planet so as to make its peculiar position in space, relative to the latter, in which it persists irrespective of distance, more or less stable. In explanation of this, suppose that one of the balls as M in **Fig. 5**[1] is not of homogenous material and that it is similarly supported but on an axis passing through its center of gravity instead of form. Then, no matter in what position the ball is fixed on the pivots, its kinetic energy and centrifugal pull will be the same. Nevertheless a directive tendency will exist as the two centers do not coincide and there is, consequently, no *dynamic* balance. When permitted to turn freely on the axis of gravity the body, of whatever shape it may be, will tend to place itself so that the line joining the two centers points to O and there may be two positions of stability but, generally, if the center of gravity is not greatly displaced, the heavier side will swing outwardly. Such condition may obtain in the moon if it had solidified before receding from the earth to great distance, when the arrangement of the masses in its interior became subject to gravitational forces of its own, vastly greater than the terrestrial. It has been suggested that the planet is egg-shaped or ellipsoidal but the departure from spherical form must be inconsiderable. It may even be a perfect sphere with the centers of gravity and symmetry coinciding and still rotate as it does. Whatever be its origin and past

---
[1] p. 175

The Moon's Rotation (follow-up) 183

history, the fact is, that at present all its parts have the same angular velocity as though it were rigidly connected with the earth. This state must endure forever unless forces from without the luna-terrestrial system bring about different conditions and thus the hope of the stargazers that its other side may become visible some day must be indefinitely deferred.

A motion of this character, as I have shown, precludes the possibility of axial rotation. The easiest way to free ourselves of this illusion is to conceive the satellite subdivided into minute and entirely independent parts, as dust particles, which have different orbital, but rigorously the same angular, velocities. One must at once recognize that the kinetic energy of such an agglomeration[2] is solely translational, there being absolutely no tendency to axial rotation. This makes it also perfectly clear why the moon, provided its distance does not greatly increase, must always turn the same face to us even *without any inherent directive tendency* nor so much as *the slightest effort from the earth*.

Referring to the librations of longitude, I do not see that they have any bearing on this question. In astronomical treatises the axial rotation of the moon is accepted as a material fact and it is thought that its angular velocity is constant while that of the orbital movement is not, this resulting in an apparent oscillation revealing more of its surface to our view. To a degree this may be true, but I hold that the mere change of orbital velocity, as will be evident from what has been stated before could not produce these phenomena, for no matter how fast or slow the gyration, the position of the body relative to the center of attraction remains the same. The real cause of these axial displacements is the changing distance of the moon from the earth owing to which the tangential components of velocity of its parts are varied. In *apogee*,[3] when the planet recedes,

[2] The state of being gathered into a mass.
[3] In astronomy, the point in the orbit of the moon at which it is furthest from earth.

the radial component of velocity decreases while the tangential increases but, as the decrement[4] of the former is the same for all parts, this is more pronounced in the regions facing the earth than in those turned away from it, the consequence being an axial displacement exposing more of the eastern side. In *perigee*,[5] on the contrary, the radial component increases and the effect is just the opposite with the result that more of the western side is seen. The moon actually swings on the axis passing through its center of gravity on which it is supported like a ball on a string. The forces involved in these pendular movements are incomparably smaller than those required to effect changes in orbital velocity. If we estimate the radius of gyration of the satellite at 600 miles and its mean distance from the earth at 240,000 miles, then the energy necessary to rotate it once in a month would be only $(600/240,000)^2 = 1/160,000$ of the kinetic energy of the orbital movement.

---

[4] The amount by which something decreases or becomes gradually less.
[5] In astronomy, the point in the orbit of the moon at which it is closest to earth.

Mr. Tesla makes a very important contribution to the electrical arts with this article.

The pioneer of all high frequency apparatus divulges much that is new and startling in these pages. Few people realize the enormous value of Mr. Tesla's machines and the many different important uses to which they can be applied in our everyday lives. New and startling uses are being found every year for these machines.

It is characteristic of Mr. Tesla that he has developed and actually built an astounding variation of these machines, and we regret that we can publish only a very few of the more important models.

Most of the Tesla coils shown have never been published before.

—Editor.
Electrical Experimenter
July 1919

Electrical Oscillators

Few fields have been opened up the exploration of which has proved as fruitful as that of high frequency currents. Their singular properties and the spectacular character of the phenomena they presented immediately commanded universal attention. Scientific men became interested in their investigation, engineers were attracted by their commercial possibilities, and physicians recognized in them a long-sought means for effective treatment of bodily ills. Since the publication of my first researches in 1891, hundreds of volumes have been written on the subject and many invaluable results obtained through the medium of this new agency. Yet, the art is only in its infancy and the future has incomparably bigger things in store.

From the very beginning I felt the necessity of producing efficient apparatus to meet a rapidly growing demand and during the eight years succeeding my original announcements I developed not less than fifty types of these transformers or electrical oscillators, each complete in every detail and refined to such a degree that I could not materially improve any one of them today. Had I been guided by practical considerations I might have built up an immense and profitable business, incidentally rendering important services to the world. But the force of circumstances and the ever enlarging vista of greater achievements turned my efforts in other directions. And so it comes that instruments will shortly be placed on the market which, oddly enough, were perfected twenty years ago!

These oscillators are expressly intended to operate on direct and alternating lighting circuits and to generate damped and undamped oscillations or currents of any frequency, volume, and tension within the widest limits. They are compact, self-contained, require no care for long periods of time, and will be found very convenient and useful for various purposes as, wireless telegraphy and telephony; conversion of electrical energy; formation of chemical compounds through fusion and combination; synthesis of gases; manufacture of ozone;[1] lighting; welding; municipal, hospital, and domestic sanitation and sterilization, and numerous other applications in scientific laboratories and industrial institutions. While these transformers have never been described before, the general principles underlying them were fully set forth in my published articles and patents, more particularly those of September 22, 1896,[2] and it is thought, therefore, that the appended photographs of a few types, together with a short explanation, will convey all the information that may be desired.

The essential parts of such an oscillator are: a condenser, a self-induction coil for charging the same to a high potential, a circuit controller, and a transformer which is energized by the oscillatory discharges of the condenser. There are at least three, but usually four, five, or six circuits in tune and the regulation is effected in several ways, most frequently merely by means of an adjusting screw. Under favorable conditions an efficiency as high as 85% is attainable, that is to say, that percentage of the energy supplied can be recovered in the secondary of the transformer. While the chief virtue of this kind of apparatus is obviously due to the wonderful

---

[1] A colorless unstable toxic gas with a pungent odor and powerful oxidizing properties, formed from oxygen by electrical discharges or ultraviolet light.
[2] Mentioned on pp. 149–150

Electrical Oscillators

powers of the condenser, special qualities result from concatenation of circuits under observance of accurate harmonic relations, and minimization of frictional and other losses which has been one of the principal objects of the design.

Broadly, the instruments can be divided into two classes: one in which the circuit controller comprises solid contacts, and the other in which the make and break is effected by mercury. **Figs. 1-8**, inclusive, belong to the first, and the remaining ones to the second class. The former are capable of an appreciably higher efficiency on account of the fact that the losses involved in the make and break are reduced to the minimum and the resistance component of the damping factor is very small. The latter are preferable for purposes requiring larger output and a great number of breaks per second. The operation of the motor and circuit controller of course consumes a certain amount of energy which, however, is the less significant the larger the capacity of the machine.

In **Fig. 1** is shown one of the earliest forms of oscillator constructed for experimental purposes. The condenser

Fig. 1 — Oscillator with detachable transformer for experimental purposes.

is contained in a square box of mahogany upon which is mounted the self-induction or charging coil wound, as will be noted, in two sections connected in multiple or series according to whether the tension of the supply circuit is 110 or 220 volts. From the box protrude four brass columns carrying a plate with the spring contacts and adjusting screws as well as two massive terminals for the reception of the primary of the transformer. Two of the columns serve as condenser connections while the other pair is employed to join the binding posts of the switch in front to the self-inductance and condenser. The primary coil consists of a few turns of copper ribbon to the ends of which are soldered short rods fitting into the terminals referred to. The secondary is made in two parts, wound in a manner to reduce as much as possible the distributed capacity and at the same time enable the coil to withstand a very high pressure between its terminals at the center, which are connected to binding posts on two rubber columns projecting from the primary. The circuit connections may be slightly varied but ordinarily they are as diagrammatically illustrated in the *Electrical Experimenter* for May,[3] relating to my oscillation transformer photograph of which appeared in the same number.[4] The operation is as follows: When the switch is thrown on, the current from the supply circuit rushes through the self-induction coil, magnetizing the iron core within and separating the contacts of the controller. The high tension induced current then charges the condenser and upon closure of the contacts the accumulated energy is released through the primary, giving rise to a long series of oscillations which excite the tuned secondary circuit.

  This device has proved highly serviceable in carrying on laboratory experiments of all kinds. For instance, in studying phenomena of impedance, the transformer was removed and a bent copper

---
[3] See p. 59
[4] See p. 56

bar inserted in the terminals. The latter was often replaced by a large circular loop to exhibit inductive effects at a distance or to excite resonant circuits used in various investigations and measurements. A transformer suitable for any desired performance could be readily improvised and attached to the terminals and in this way much time and labor was saved. Contrary to what might be naturally expected, little trouble was experienced with the contacts, although the currents through them were heavy, namely, proper conditions of resonance existing, the great flow occurs only when the circuit is closed and no destructive arcs can develop. Originally I employed platinum and iridium tips but later replaced them by some of meteorite and finally of tungsten. The last have given the best satisfaction, permitting working for hours and days without interruption.

Fig. 2 illustrates a small oscillator designed for certain specific uses. The underlying idea was to attain great activities during minute intervals of time each succeeded by a comparatively long period of inaction. With this object a large self-induction and a quick-acting break were employed owing to which arrangement the condenser was charged to a very high potential. Sudden secondary currents and sparks of great volume were thus obtained, eminently suitable for welding thin wires, flashing lamp filaments, igniting explosive mixtures and kindred applications. The instrument was also adapted for battery use

Fig. 2 — Small Tesla Coil for gas engine ignition and similar uses.

and in this form was a very effective igniter for gas engines on which a patent bearing number 609,250 was granted to me August 16, 1898.

Fig. 3 represents a large oscillator of the first class intended for wireless experiments, production of Röntgen rays[5] and scientific research in general. It comprises a box containing two condensers of the same capacity on which are supported the charging coil and transformer. The automatic circuit controller, hand switch, and connecting posts are mounted on the front plate of the inductance spool as is also one of the contact springs. The condenser box is equipped with three terminals, the two external ones serving merely for connection while the middle one carries a contact bar with a screw for regulating the interval during which the circuit is closed. The vibrating spring, itself, the sole function of which is to

Fig. 3 — Tesla Transformer, 12-inch spark, chiefly for wireless.

cause periodic interruptions, can be adjusted in its strength as well as distance from the iron core in the center of the charging coil by four screws visible on the top plate so that any desired conditions of mechanical control might be secured. The primary coil of the transformer is of copper sheet and taps are made at suitable points for the purpose of varying, at will, the number of turns. As in **Fig. 1** the inductance coil is wound in two sections to adapt the instrument

---

[5] X-rays. Sometimes referred to as Röntgen radiation, after the German scientist Wilhelm Conrad Röntgen (1845–1923) who discovered it.

both to 110 and 220 volt circuits and several secondaries were provided to suit the various wave lengths of the primary. The output was approximately 500 watt with damped waves of about 50,000 cycles per second. For short periods of time undamped oscillations were produced in screwing the vibrating spring tight against the iron core and separating the contacts by the adjusting screw *which also performed the function of a key.* With this oscillator I made a number of important observations and it was one of the machines exhibited at a lecture before the New York Academy of Sciences in 1897.[6]

Fig. 4 is a photograph of a type of transformer in every respect similar to the one illustrated in the May, 1919, issue of the *Electrical Experimenter*[7] to which reference has already been made. It contains the identical essential parts, disposed in like manner, but was specially designed for use on supply circuits of higher tension, from 220 to 500 volts or more. The usual adjustments are made in setting the contact spring and shifting the

Fig. 4 — Tesla Oscillator in action, generating undamped waves.

[6] Titled "The Streams of Lenard and Roentgen and Novel Apparatus for Their Production."
[7] See p. 59 (again).

iron core within the inductance coil up and down by means of two screws. In order to prevent injury through a short-circuit, fuses are inserted in the lines. The instrument was photographed in action, generating undamped oscillations from a 220 volt lighting circuit.

Fig. 5 shows a later form of transformer principally intended to replace Rhumkorf coils. In this instance a primary is employed, having a much greater number of turns and the secondary is closely linked with the same. The currents developed in the latter, having a tension of from 10,000 to 30,000 volts, are used to charge condensers and operate an independent high frequency coil as customary. The controlling mechanism is of somewhat different construction but the core and contact spring are both adjustable as before.

Fig. 5 — Later type of Tesla Transformer.

Fig. 6 is a small instrument of this type, particularly intended for ozone production or sterilization. It is remarkably efficient for its size and can be connected either to a 110 or 220 volt

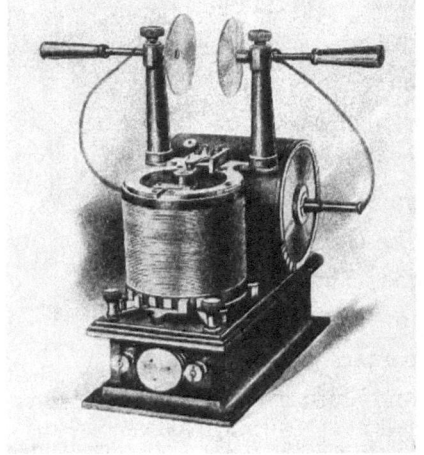

Fig. 6 — Small oscillator for production of ozone.

circuit, direct or alternating, preferably the former.

In **Fig. 7** is shown a photograph of a larger transformer of this kind. The construction and disposition of the parts is as before but there are two condensers in the box, one of which is connected in the circuit as in the previous cases, while the other is in shunt to the primary coil. In this manner currents of great volume are produced in the latter and the secondary effects are accordingly magnified. The introduction of an additional tuned circuit secures also other advantages but the adjustments are rendered more difficult and for this reason it is desirable to use such an instrument in the production of currents of a definite and unchanging frequency.

Fig. 7 — Large Tesla Transformer for various purposes.

Fig. 8 illustrates a transformer with rotary break. There are two condensers of the same capacity in the box which can be connected in series or multiple. The charging inductances are in the form of two long spools upon which are supported the secondary terminals. A small direct current motor, the speed of which can be varied within wide limits, is employed to drive a specially constructed make and break.

Fig. 8 — Tesla Transformer with rotary break for wireless.

In other features the oscillator is like the one illustrated in **Fig. 3** and its operation will be readily understood from the foregoing. This transformer was used in my wireless experiments and frequently also for lighting the laboratory by my vacuum tubes and was likewise exhibited at my lecture before the New York Academy of Sciences previously mentioned.

Coming now to machines of the second class, **Fig. 9** shows an oscillatory transformer comprising a condenser and charging inductance enclosed in a box, a transformer, and a mercury circuit controller, the latter being of a construction described for the first time in my patent No. 609,251 of August 16, 1898. It consists of a motor driven hollow pulley containing a small quantity of mercury which is thrown outwardly against the walls of the vessel by centrifugal force and entrains a contact wheel which periodically closes and opens the condenser circuit. By means of adjusting screws above the pulley, the depth of immersion of the vanes and consequently,

Fig. 9 — Tesla Transformer with mercury interrupter.

also, the duration of each contact can be varied at desire and thus the intensity of the effects and their character controlled. This form of break has given thorough satisfaction, working continuously with currents of from 20 to 25 amperes. The number of interruptions is usually from 500 to 1,000 per second but higher frequencies are practicable. The space occupied is about 10" x 8" x 10" and the output approximately ½ kW.

In the transformer just described the break is exposed to the atmosphere and a slow oxidation of the mercury takes place. This disadvantage is overcome in the instrument shown in **Fig. 10**, which consists of a perforated metal box containing the condenser and charging inductance and carrying on the top a motor driving the break, and a transformer. The mercury break is of a kind to be described and operates on the principle of a jet which establishes, intermittently, contact with a rotating wheel in the interior of the pulley. The stationary parts are supported in the vessel on a bar passing through the long hollow shaft of the motor and a mercury seal is employed to effect hermetic closure of the chamber enclosing the circuit controller. The current is led into the interior of the pulley through two sliding rings on the top which are

Fig. 10 — Large Tesla Transformer with hermetically sealed mercury interrupter.

Fig. 11 — Tesla Transformer with sealed mercury interrupter for low tension work.

in series with the condenser and primary. The exclusion of the oxygen is a decided improvement, the deterioration of the metal and attendant trouble being eliminated and perfect working conditions continuously maintained.

Fig. 11 is a photograph of a similar oscillator with hermetically enclosed mercury break. In this machine the stationary parts of the interrupter in the interior of the pulley were supported on a tube through which was led an insulated wire connecting to one terminal of the break while the other was in contact with the vessel. The sliding rings were, in this manner, avoided and the construction simplified. The instrument was designed for oscillations of lower tension and frequency requiring primary currents of comparatively smaller amperage and was used to excite other resonant circuits.

Fig. 12 shows an improved form of oscillator of the kind described in Fig. 10, in which the supporting bar through the hollow motor shaft was done away with, the device pumping the mercury being kept in position by gravity, as will be more fully explained with reference to another figure. Both the capacity of the condenser and primary turns were made variable with the view of producing oscillations of several frequencies.

Fig. 13 is a photographic view of another form of oscillatory transformer with hermetically sealed mercury interrupter, and Fig. 14 diagrams showing the circuit connections and arrangement of parts reproduced from my patent, No. 609,245, of August 16, 1898, describing this particular device. The condenser, inductance, transformer and circuit controller are disposed as before, but the latter is of different construction, which will be clear from an inspection of Fig. 14. The hollow pulley a is secured to a shaft c which is mounted in a vertical bearing passing through the stationary field magnet d of the motor. In the interior of the vessel is supported, on frictionless bearings, a body h of magnetic material which is surrounded by a dome b in the center of a laminated iron ring, with pole pieces oo wound

Fig. 12 — Another type of Tesla Transformer with mercury interrupter.

Fig. 13 — Tesla Transformer with rotary break for wireless.

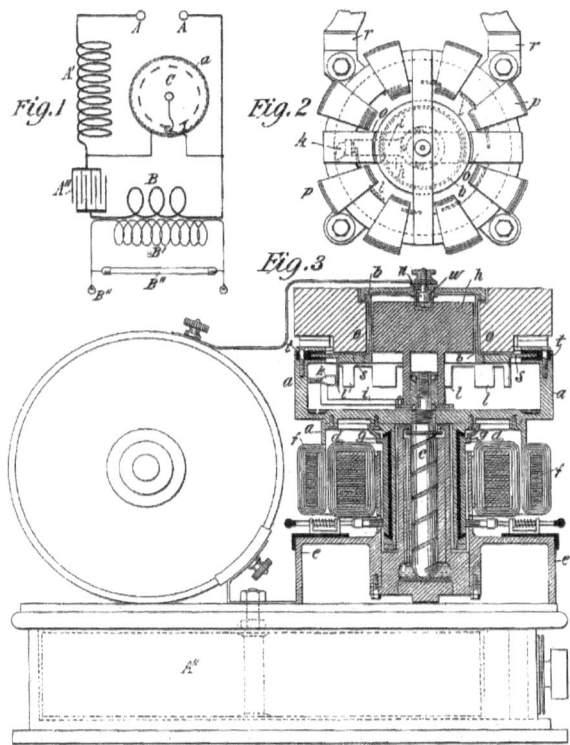

Fig. 14 — Electrical Oscillator, illustrated in **Fig. 13**, showing details and circuit connections.

with energizing coils **p**. The ring is supported on four columns and, when magnetized, keeps the body **h** in position while the pulley is rotated. The latter is of steel, but the dome is preferably made of German silver burnt black by acid or nickeled. The body **h** carries a short tube **k** bent, as indicated, to catch the fluid as it is whirled around, and project it against the teeth of a wheel fastened to the pulley. The wheel is insulated and contact from it to the external circuit is established through a mercury cup. As the pulley is rapidly rotated a jet of the fluid is thrown against the wheel, *thus making and breaking contact about 1,000 times per second*. The instrument works silently and, owing to the absence of all deteriorating agents, keeps continually clean and in perfect condition. The number of interruptions per second may be much greater, however, so as to make the currents suitable for wireless telephony and like purposes.

Fig. 15 — Tesla Transformer with gravity controlled, sealed mercury interrupter.

A modified form of oscillator is represented in **Figs. 15** and **16**, the former being a photographic view and the latter a diagrammatic illustration showing the arrangement of the interior parts of the controller. In this instance the shaft **B** carrying the vessel **A** is hollow and supports, in frictionless bearings, a spindle **j** to which is fastened a weight **K**. Insulated from the latter, but mechanically fixed to it, is a curved arm **L** upon which is supported, freely rotatable, a break-wheel with projections **QQ**. The wheel is in electrical connection

Electrical Oscillators 201

with the external circuit through a mercury cup and an insulated plug supported from the top of the pulley. Owing to the inclined position of the motor the weight K keeps the break-wheel in place by the force of gravity and as the pulley is rotated the circuit, including

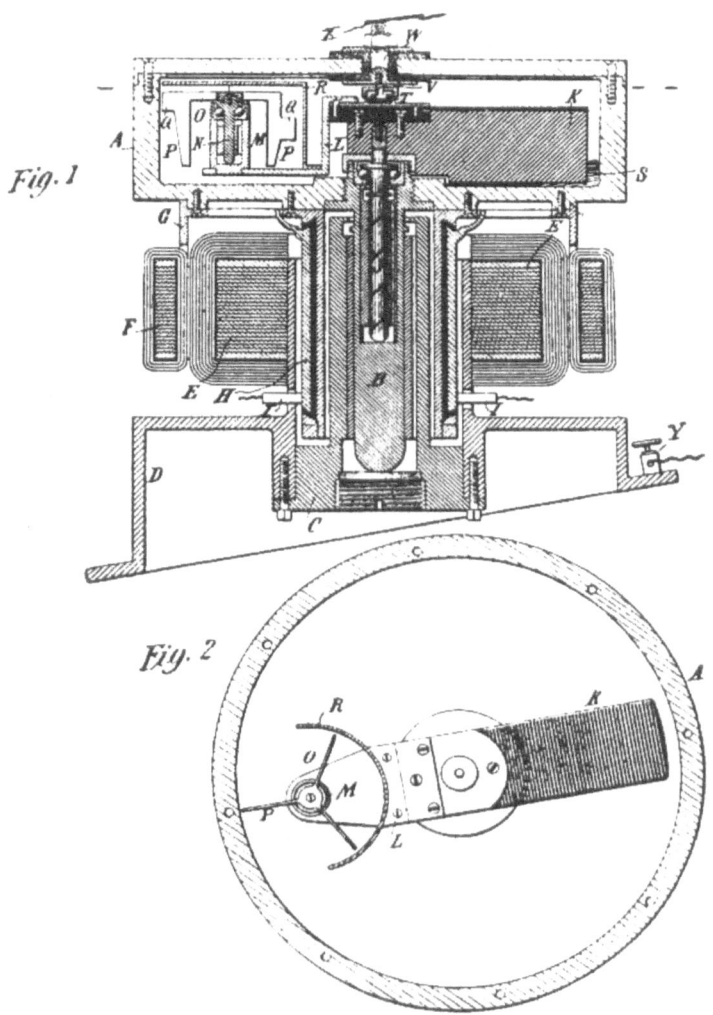

Fig. 16 — Electrical Oscillator, illustrated in **Fig. 15**, showing details of motor and break mechanism.

the condenser and primary coil of the transformer, is rapidly made and broken.

Fig. 17 shows a similar instrument in which, however, the make and break device is a jet of mercury impinging against an insulated toothed wheel carried on an insulated stud in the center of the cover of the pulley as shown. Connection to the condenser circuit is made by brushes bearing on this plug.

Fig. 18 is a photograph of another transformer with a mercury circuit controller of the wheel type, modified in some features on which it is unnecessary to dwell.

These are but a few of the oscillatory transformers I have perfected and constitute only a small part of my high frequency apparatus of which I hope to give a full description, when I shall have freed myself of pressing duties, at some future date.

Fig. 17 — Tesla Transformer with gravity controlled, sealed mercury interrupter.

Fig. 18 — Tesla Transformer with mercury jet interrupter.

+ appendices

Tesla Has New Pointless Lightning Rod

Since the introduction of the lightning rod over one hundred years ago by Benjamin Franklin, its adoption as a means of protection against destructive atmospheric discharges such as lightning bolts, has been practically universal. In a recent discussion on the subject of lightning protection, Dr. Nikola Tesla of New York, brings out many interesting facts not generally known concerning the real efficacy of the ordinary lightning rod as installed on houses, barns, and public buildings all over the world.

Says, Dr. Tesla, "The efficacy of the ordinary lightning rod is to a certain degree unquestionably established through statistical records, but there is generally a prevalent, nevertheless, a singular theoretical fallacy as to its operation, and its construction is radically defective in one feature, namely its typical pointed terminal." In his new form of lightning protecting rod and terminal here illustrated, Tesla avoids all such points on the metal parts facing skyward, and uses an entirely different form and arrangement of terminals.

In permitting leakage into the air, the needle-shaped lightning rod is popularly believed to perform two functions: one to drain the ground of its negative electricity, the other to neutralize the positive electricity of the clouds. To some degree it does both. But a systematic study of electrical disturbances in the earth has made it palpably evident that the action of Franklin's conductor, as commonly interpreted, is chiefly illusionary. Actual measurement proves the quantity of electricity escaping even from many points, to be entirely insignificant when compared with that induced within a considerable terrestrial area, and of no moment whatever in the process of dissipation. But it is true that the negatively charged air in the vicinity of the rod, rendered conductive through the influence of the same, facilities the passage of the bolt. Therefore it increases the probability of a lightning discharge in the vicinity. The fundamental facts underlying this type of lightning-rod are: First, it attracts lightning, so that it will be struck more often than would be the building if it were not present; second, it renders harmless most, *but not all*, of the discharges which it receives; third, by rendering the air conductive, and for other reasons, it is sometimes the cause of damage to neighboring objects; and fourth, on the whole, its power of preventing injury predominates, more or less, over the hazards it invites.

By contrast, Tesla's new lightning protector is founded on principles diametrically opposite. Its terminal has a *large surface*. It secures a very *low density* and preserves the insulating qualities of the ambient medium, thereby minimizing leakage, and thus acting as a quasi-repellent to increase enormously the safety factor.

An understanding of but part of the truths relative to electrical discharges, and their misapplication due to the want of fuller appreciation has doubtless been responsible for the Franklin lightning rod taking its conventional *pointed form*, but theoretical considerations, and the important discoveries that have been made in the course of investigations with a Tesla wireless transmitter of great activity by which arcs of a volume and tension actually comparable to those occurring in nature were obtained, at once establish the fallacy of the hitherto prevailing notion on which the Franklin type of rod is based and show the distinctive novelty of this new lightning protector.

Practical estimates of the electrical quantities concerned in natural disturbances show, moreover, how absolutely impossible are the functions attributed to the pointed lightning conductor. A single cloud may contain several billion electric units, or more, inducing in the earth an equivalent amount, which a number of lightning rods *could not neutralize in many years*. Particularly to instance conditions that may have to be met, reference is made to an actual case (in 1904) wherein it appears that upon one occasion approximately 12,000 strikes occurred within two hours, all within a radius of less than 31 miles from the place of observation.

But although the pointed lightning rod is quite ineffective in the one respect noted, it has the property of attracting lightning to a high degree — first, on account of its shape and secondly, because it ionizes and renders conductive the surrounding air. This has been unquestionably established in long continued tests with the Tesla wireless transmitter already mentioned, the inventor claims, and in this feature lies the chief disadvantage of the Franklin type of protector.

In **Fig. A** and **Fig. B** on the following page, different forms of such low density terminals and the arrangement of the same are illustrated. In **Fig. A**, there is a cast or spun metal shell of ellipsoidal outline, having on its under side a sleeve with a bushing of porcelain or other insulating material, adapted to be slipped tightly on a metal rod, which may be an ordinary lightning conductor. **Fig. B** shows another form of terminal made up of rounded or flat metal bars radiating from a central hub, which is supported directly on a metal rod and in electrical contact with the same. The special object of this type is to reduce the wind resistance, but it is essential that the bars have a *sufficient area to insure small electrostatic density*, and also that they are close enough to make the aggregate capacity nearly equal to that of a continuous shell of the same outside dimensions. The general view of the building shows a cupola-shaped[1] and earthed metal dome carried by a chimney, serving in this way the twofold practical purpose of hood and protector.

From the foregoing it will be clear that in all cases the new Tesla terminal prevents leakage of electricity and attendant ionization of the air. It is immaterial to this end whether it is insulated or not. Should it be struck the current will pass readily to the ground either directly or, as in **Fig. A**, through a small air-gap between. But such an accident is rendered extremely improbable owing to the fact that there are everywhere points and projections on which the terrestrial charge attains a high density and where the air is ionized. Thus the action of the improved protector is equivalent to a repellent force. This being so, it is not necessary to support it at a great height, but the ground connection should be made with the usual care and the conductor leading to it must be of as small a self-induction and resistance as practicable. Tesla has taken out a patent on this new lightning protector.

---

[1] Dome-shaped.

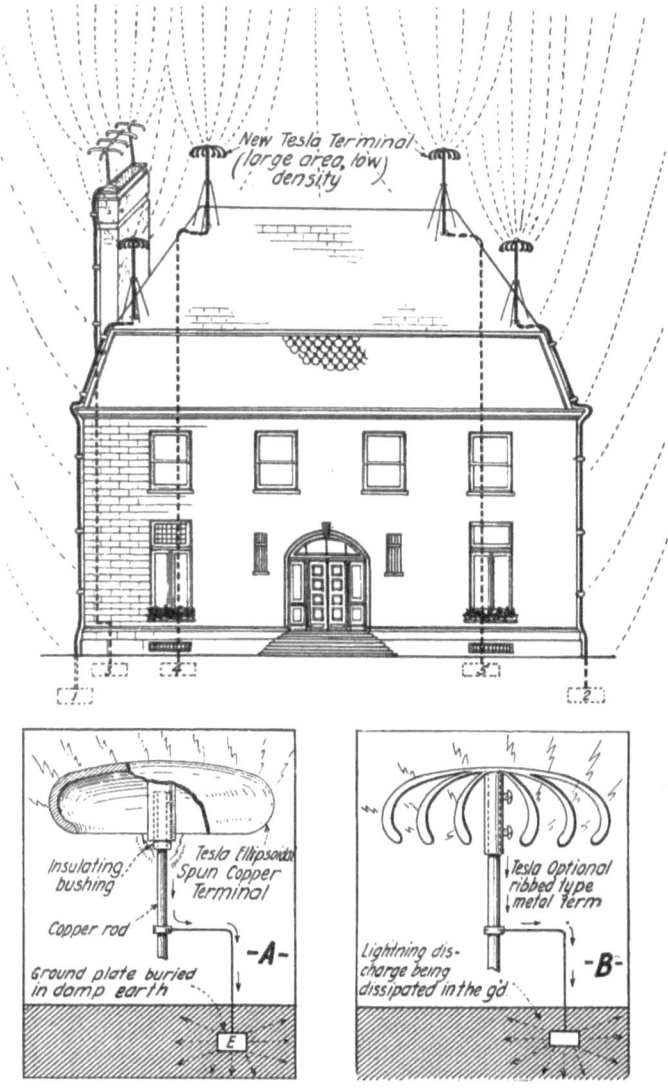

Nikola Tesla, expert on high frequency currents, such as lightning discharges, has recently patented the new "rounded" form of lightning rod, which he states is superior to the time-honored "pointed" rods so familiar to all of us. Also, Dr. Tesla has good reasons for this radical departure in lightning rod design.

Tesla's Egg of Columbus

How Tesla Performed the Feat of Columbus Without Cracking the Egg

Probably one of the most far-reaching and revolutionary discoveries made by Mr. Tesla is the so-called *rotating magnetic field*. This is a new and wonderful manifestation of force — a magnetic cyclone — producing striking phenomena which amazed the world when they were first shown by him. It results from the joint action of two or more alternating currents definitely related to one another and creating magnetic fluxes, which, by their periodic rise and fall according to a mathematical law, cause a continuous shifting of the lines of force. There is a vast difference between an ordinary electro-magnet and that invented by Tesla. In the former the lines are stationary, in the latter they are made to whirl around at a furious rate. The first attracts a piece of iron and holds it fast; the second causes it to spin in any direction and with any speed desired. Long ago, when Tesla was still a student, he conceived the idea of the rotating magnetic field and this remarkable principle is embodied in his famous *induction motor* and system of transmission of power now in universal use.

MY INVENTIONS & other essays

In this issue of the *Electrical Experimenter* Mr. Tesla gives a remarkable account of his early efforts and trials as an inventor and of his final success.[1] Unlike other technical advances arrived at through the usual hit and miss methods and hap-hazard experimentation, the rotating field was purely the work of scientific imagination. Tesla developed and perfected, entirely in his mind, this great idea in all its details and applications *without making one single experiment*. Not even the usual first model was used. When the various forms of apparatus he had devised were tried for the first time they worked exactly as he had imagined and he took out some forty fundamental patents covering the whole vast region he had explored. He obtained the first rotations in the summer of 1883 after five years of constant and intense thought on the subject and then undertook the equally difficult task of finding believers in his discovery. The alternating current was but imperfectly understood and had no standing with engineers or electricians and for a long time Tesla talked to deaf ears. But, ultimately, his pains were rewarded and early in 1887 a company bearing his name was formed for the commercial introduction of the invention.

Dr. Tesla recently told the editors an amusing incident in this connection. He had approached a Wall Street capitalist — a prominent lawyer — with a view of getting financial support and this gentleman called in a friend of his, a well-known engineer at the head of one of the big corporations in New York, to pass upon the merits of the scheme. This man was a practical expert who knew of the failures in the industrial exploitation of alternating currents and was distinctly prejudiced to a point of not caring even to witness some tests. After several discouraging conferences Mr. Tesla had an inspiration. Everybody has heard of the "Egg of Columbus." The saying goes that at a certain dinner the great explorer asked some scoffers of his project to balance an egg on its end. They tried it in

---
[1] This article was published in the same issue of *EE* as Chapter 2: "My First Efforts at Invention."

vain. He then took it and cracking the shell slightly by a gentle blow, made it stand upright. This may be a myth but the fact is that he was granted an audience by Isabella, the Queen of Spain, and won her support. There is a suspicion that she was more impressed by his portly bearing than the prospect of his discovery. Whatever it might have been, the Queen pawned her jewels and three ships were equipped for him and so it happened that the Germans got all that was coming to them in this war. But to return to Tesla's reminiscence. He said to these men, "Do you know the story of the Egg of Columbus?" Of course they did. "Well," he continued, "what if I could make an egg stand on the pointed end without cracking the shell?"

"If you could do this we would admit that you had gone Columbus one better."

"And would you be willing to go out of your way as much as Isabella?"

"We have no crown jewels to pawn," said the lawyer, who was a wit, "but there are a few ducats[2] in our buckskins and we might help you to an extent."

Mr. Tesla thus succeeded in capturing the attention and personal interest of these very busy men, extremely conservative and reluctant to go into any new enterprise, and the rest was easy. He arranged for a demonstration the following day. A rotating field magnet was fastened under the top board of a wooden table and Mr. Tesla provided a copper-plated egg and several brass balls and pivoted iron discs for convincing his prospective associates. He placed the egg on the table and, to their astonishment, it stood on end, but when they found that it was *rapidly spinning* their stupefaction was complete. The brass balls and pivoted iron discs in turn were set spinning rapidly by the rotating field, to the amazement of

[2] Any of various gold coins formerly issued in several parts of Europe.

MY INVENTIONS & other essays

the spectators. No sooner had they regained their composure than Tesla was delighted with the question: "Do you want any money?"

"Columbus was never in a worse predicament," said the great inventor, who had parted with his last portrait of George Washington in defraying the expenses of the preparation. Before the meeting adjourned he had a substantial check in his pocket, and it was given with the assurance that there was more to be had in the same bank. That started the ball rolling. Tens of millions of horsepower of Tesla's induction motors are now in use all over the world and their production is rising like a flood.

In 1893 Mr. Albert Schmid, then Superintendent of the Westinghouse Electric and Mfg. Co. constructed a powerful rotating field ring with an egg made of copper, and larger than that of an ostrich, for Dr. Tesla's personal collection at the Chicago World's Fair. This piece of apparatus was one of the most attractive novelties ever publicly shown and drew enormous crowds every day. Subsequently it was taken to Mr. Tesla's laboratory and served there permanently for demonstrating rotating field phenomena. In his experiments it was practicable to use as much as 200 horsepower for a short time, without overheating the wires and the effects of the magnetic forces were wonderfully fascinating to observe. This is the very ring indicated in the accompanying photograph (**Fig. 1**, following page), giving a view of Mr. Tesla's former laboratory at 46 E. Houston Street, New York. It is shown in detail in **Fig. 2**, and the mode of winding is illustrated in diagram (**Fig. 3**). Originally the two-phase arrangement was provided but Mr. Tesla transformed it to the three- and four-phase when desired. On top of the ring was fastened a thin circular board, slightly hollowed, and provided around its circumference with a guard to prevent the objects from flying off.

Even more interesting than the spinning egg was the exhibition of *planetary motion*. In this experiment one large, and several small

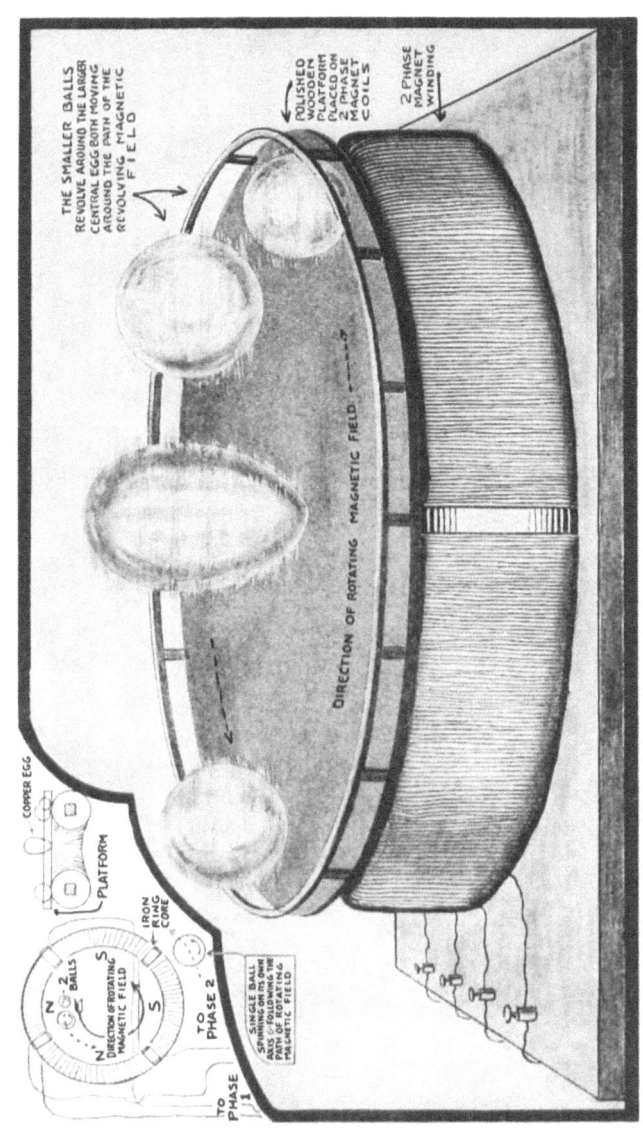

Fig. 2 — Illustrating the polyphase coil and rotating magnetic field which caused copper eggs to spin.
Fig. 3 — Insert: Detail of coil apparatus showing coil connection to different phases.

Fig. 1 — This hitherto unpublished photograph is extremely interesting as it shows not only "Tesla's Electric Egg" apparatus in the center of the background (white arrow), but also a comprehensive view of a corner of his famous Houston Street laboratory of a decade ago. At the left may be seen a number of Tesla's oscillators or high frequency generators, while in the rear may be noted a large high frequency transformer of the spiral type, the diameter of which was a little over nine feet. The electric egg apparatus comprising a two-phase A.C. circular core and winding, rests on a table, and this particular model measured about two feet across. In making the demonstrations, Tesla applied as much as 200 hp from a two-phase alternator to the exciting coils, and so intense was the revolving magnetic field created in the surrounding space, that small delicately pivoted iron discs would revolve in any part of the hall, and a great many other devices could be simultaneously operated from this magnetic field when this excited. The frequency of the two-phase A.C. energizing the coils, was varied from 25 to 300 cycles, the best results being obtained with current of from 35 to 40 cycles. This laboratory was lighted by Tesla's vacuum tubes, several of which may be seen on the ceiling, and each of which emitted 50 C.P. The coil resting on three legs and observed in the immediate foreground is the primary of a resonant Tesla transformer which collected energy from an oscillatory circuit encircling the laboratory, no matter in what position the transformer was placed. A low tension secondary of one or two turns of heavy cable (not visible) was provided for stepping down the energy collected by "mutual induction," and supplied the current to incandescent lamps, vacuum tubes, motors, and other devices. When the circuit around the hall was strongly excited, the secondary furnished energy at the rate of about three-quarters of one horsepower.

brass balls were usually employed. When the field was energized all the balls would be set spinning, the large one remaining in the center while the small ones revolved around it, like moons about a planet, gradually receding until they reached the outer guard and raced along the same.

But the demonstration which most impressed the audiences was the simultaneous operation of numerous balls, pivoted discs, and other devices placed in all sorts of positions and *at considerable distances from the rotating field*. When the currents were turned on and the whole animated with motion, it presented an unforgettable

Fig. 4 — This photograph represents a collection of a few of Tesla's wireless lamps, such as he proposes to use in lighting isolated dwellings all over the world from central wireless plants. The two lamps at either corner at the bottom are illuminated, owing to the fact that a high frequency oscillator was in operation some distance away when this photograph was taken. These tubes were filled with various gases for experimental research work in determining which was efficient.

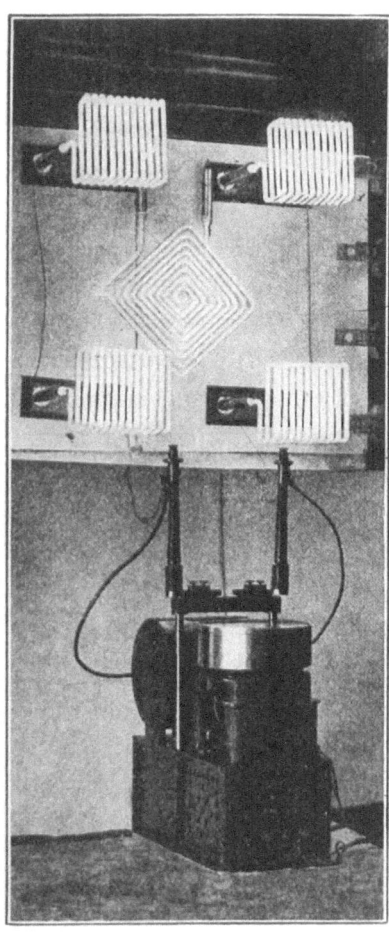

Fig. 5 — This illustration shows one of Tesla's high frequency oscillation generators and a bank of his high frequency lamps lighted by the same. These highly evacuated, gas-filled tubes were operated in different ways. In some cases they were connected to one wire only; in other instances to two wires, in the manner of ordinary incandescent lamps. Often, however, they were operated without any connection to wires at all, i.e., by "wireless energy" over quite appreciable distances, which could have been greatly extended with more power. The oscillator comprises a Tesla high potential transformer which is excited from a condenser and circuit controller, as described in his patents of 1896. The primary exciting element comprised a powerful electro-magnet actuating an armature, and this circuit was connected with 110 volt, 60 cycle A.C. or D.C. When the oscillator was put into operation, the interrupter actuated by the electro-magnet connected to the 100 volt circuit became simultaneously the spark gap for the high potential exciting circuit, which included this vibrator, spark gap, a high tension condenser, and the primary of the high frequency Tesla transformer. The lamps were connected to the secondary of the latter, the terminals of which are seen in the rear of the machine.

spectacle. Mr. Tesla had many vacuum bulbs in which small, light metal discs were pivotally arranged on jewels and these would spin anywhere in the hall when the iron ring was energized.

Rotating fields of 15,000 horsepower are now being turned out by the leading manufacturers and it is very likely that in the near future capacities of 50,000 horsepower will be employed in the steel and other industries and ship propulsion by Tesla's electric drive which, according to Secretary of the Navy Daniels' statement, has proved a great success.

But any student interested in these phenomena can repeat all the classical experiments of Tesla by inexpensive apparatus. For this purpose it is only necessary to make two slip ring connections on an ordinary small direct current motor or dynamo and to wind an iron ring with four coils as indicated in diagram **Fig. 3**. No particular rule need be given for the windings but it may be stated that he will get the best results if he will use an iron ring of comparatively *small section* and wind it *with as many turns of stout wire as practicable*. He can heavily copper plate an egg but he should bear in mind that Tesla's egg is not as innocent as that of Columbus. The worst that can happen with the latter is that it might be — er — over ripe! but the Tesla egg may explode with disastrous effect because the copper plating is apt to be brought to a high temperature through the induced currents. The sensible experimenter will, therefore, first suck out the contents of the egg — thus satisfying both his appetite and thirst for knowledge.

Besides the rotating field apparatus Mr. Tesla had other surprises for his audiences, which were even more wonderful. So, for instance, the coil on three legs, visible in the foreground, was used to operate wireless motors, lamps, and other devices, and the spiral coil in the background served to show extraordinary high potential phenomena, as streamers of great length.

As I review the events
of my past life
I realize how subtle
are the influences that
shape our destinies.

www.ingramcontent.com/pod-product-compliance
Lightning Source LLC
Chambersburg PA
CBHW030321100526
44592CB00010B/513